International Environmental Labelling

Vol.6 of 11

For All People who wish to take care of Climate Change
Wood & Stationery Industries: (Wooden Products, Cardboard,
Papers, Markers, Pens, NoteBooks. Writing Pads and Writing
Sets, Pencils, White Papers, Envelopes and Organizers, Staplers
and Paper Clips)

Jahangir Asadi

Vancouver, BC CANADA

Suggest an ecolabel

If you think that we missed a label and/or you are an ecolabel-ling body, please consider to submit for the next editions of our 11 Volumes International Eco-labelling Book series. Please send your details, and we'll review your suggestions. Our goal is to be as comprehensive as possible, so thank you for your help!
info@TopTenAward.Net

Published by: Top Ten Award International Network
Vancouver, BC **CANADA**
Email: Info@TopTenAward.net
www.TopTenAward.net

Ordering Information:
Quantity sales. Special discounts are available on quantity purchases by universities, schools, corporations, associations, and others. For details, contact the "Sales Department" at the above mentioned email address.

International Environmental Labelling Vol.6/J.Asadi—2nd ed.
ISBN 978-1-7773356-8-7

Contents

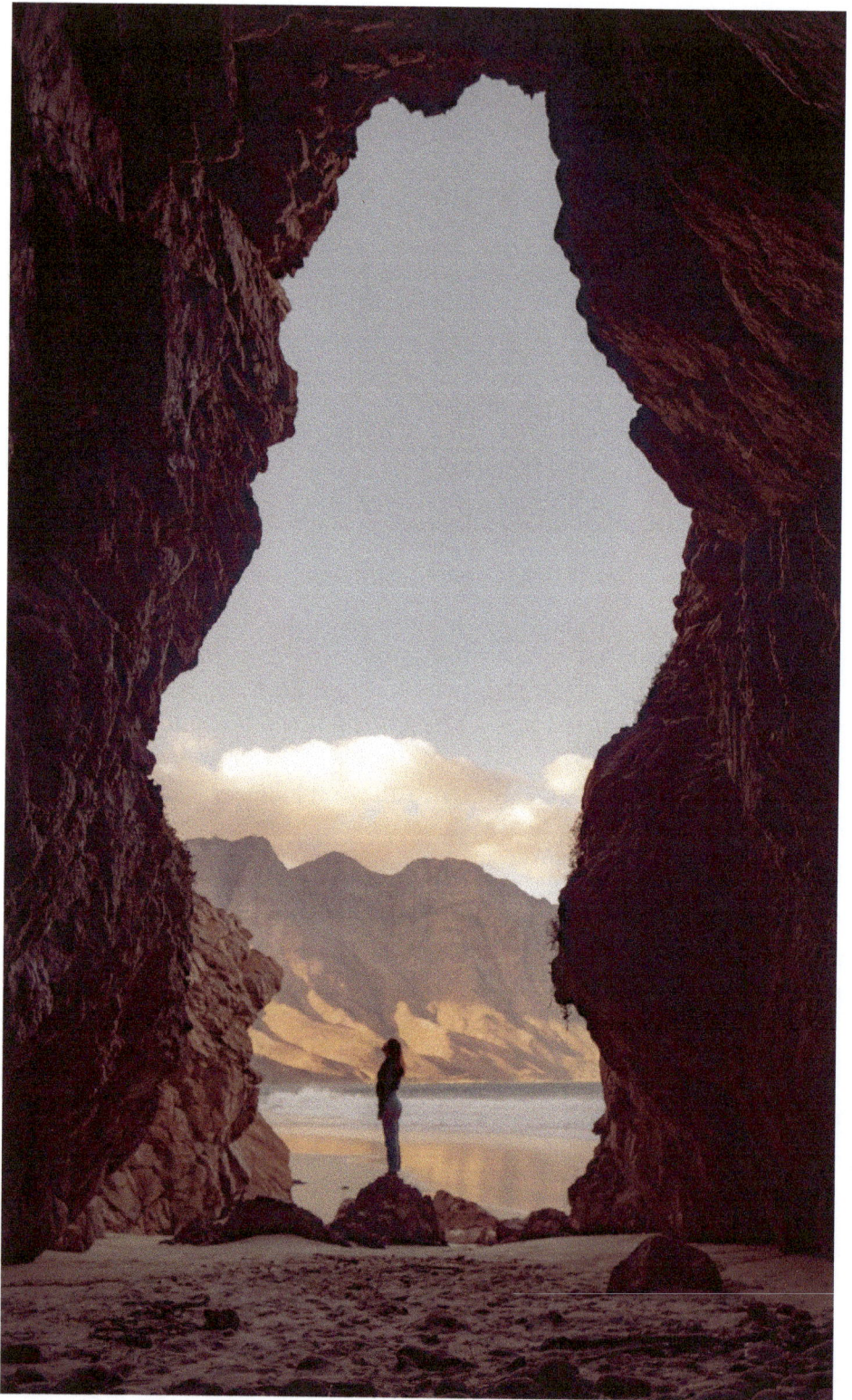

I dedicate this book to my father in law, Rasoul

We hope that, 10,000 years from now, future generations will be able to see flowers that provide bees with nectar and pollen and... BEES provide flowers with the means to reproduce by spreading pollen from flower to flower...

Jahangir Asadi

Acknowledgements:

I wish to thank my committee members, who were more than generous with their expertise and precious time. I would like to acknowledge and thank the Top Ten Award International Network for allowing me to conduct my research and providing any assistance requested.

It should be noted that all the required permissions for using the logos and trade marks has been obtained to be published in this volume.

Why do we need to use environment friendly?
Since the world is corrupted with pollution and toxic amount of materials, making it sustain- able can be a good call.
Going eco-friendly also improves your quality of life in terms of mortality, age, diseases etc.
You might have a better shot at living a quality life with health if you chose to go eco-friendly

Top Ten Award International Network

Top Ten Award international Network (TTAIN) was established in 2012 to recognize outstanding individuals, groups, companies, organizations representing the best in the public works profession.

TTAIN publishing books related to international Eco-labeling plans to increase public knowledge in purchasing based on the environmental impacts of products.

Top Ten Award International Network provides A to Z book publishing services and distribution to over 39,000 booksellers worldwide, including Apple, Amazon, Barnes & Noble, Indigo, Google Play Books, and many more.

Our services including: editing, design, distribution, marketing

TTAIN Book publishing are in the following categories:

Student
Standard
Business
Professional
Honorary

We focus on quality, environmental & food safety management systems , as well as environmnetal sustain for future kids. TTAIN also provide complete consulting services for QMS, EMS, FSMS, HACCP and Ecolabeling based on international standards.

ISO 14024 establishes the principles and procedures for developing Type I environmental labelling programmes, including the selection of product categories, product environmental criteria and product function characteristics, and for assessing and demonstrating compliance. ISO 14024 also establishes the certification procedures for awarding the label.

TTAIN has enough experiences to help create new ecolabeling programmes in different countries all over the world.
For more detail visit our website : http://toptenaward.net
and/or send your enquiery to the following email:
info@toptenaward.net

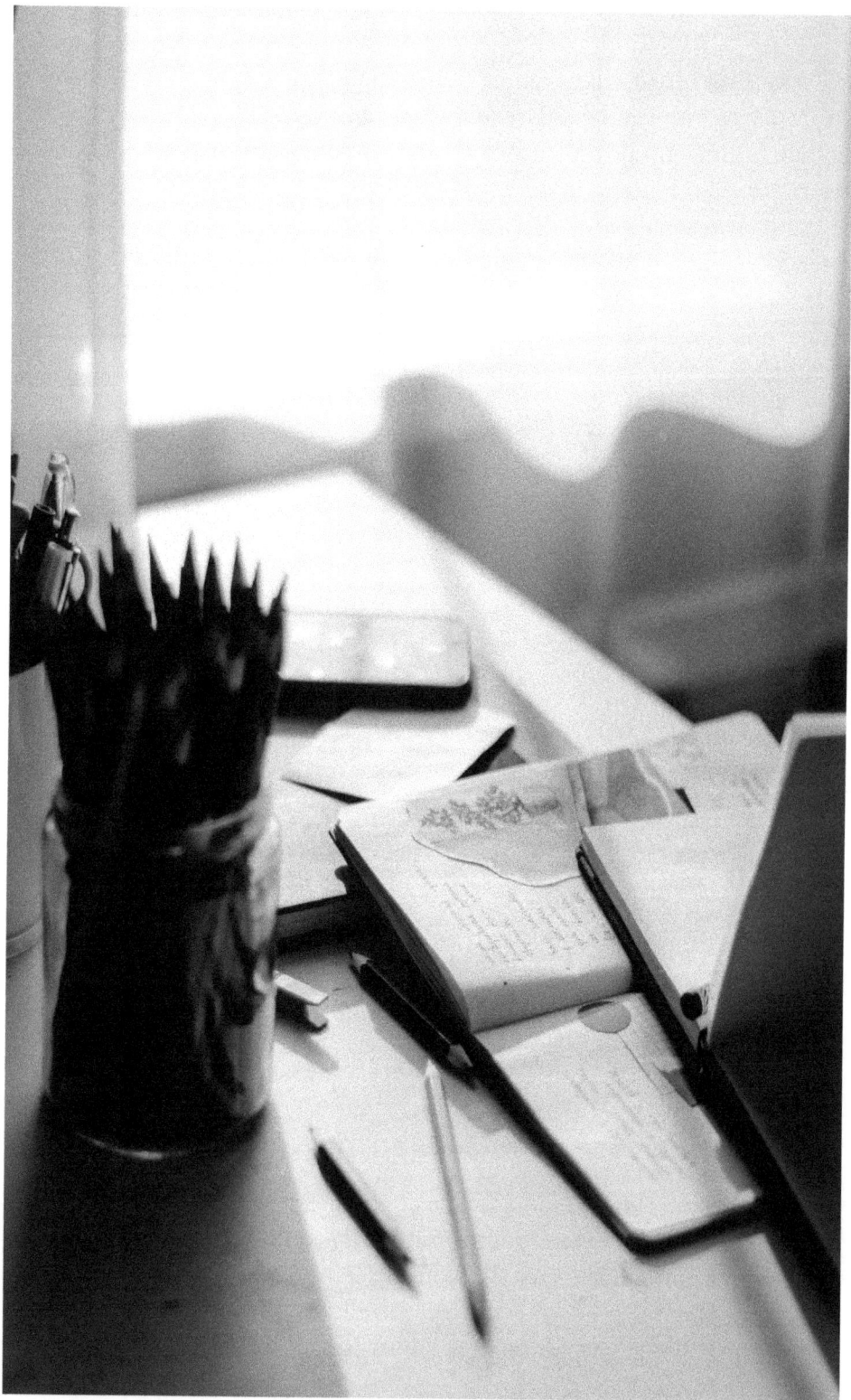

CHAPTER 1

Introduction

This book is dedicated to the subject of environmental labels. The basis for the classification of its parts goes back to the types of environmental labelling according to the classifications provided by the International Organization for Standardization. In each section, while presenting the relevant definitions, I mention the existing international standards and present examples related to each type of labelling. Environmental labelling is an important and significant topic, and its richness is added to every day, which has attracted the attention of many experts and researchers around the world. The idea of compiling this book, came to my mind when I observed that national environmental labelling models have been developed in most countries of the world, but in many other countries, the initial steps have not been taken yet. Therefore, I decided to create the first spark for the development of environmental labelling patterns in other countries by collecting appropriate materials and inserting samples of labelling patterns of different countries of the world. It should be noted that the description of each environmental label in this book does not indicate their approval or denial; they are included only to increase the awareness of all enthusiasts and consumers of the meanings and concepts derived from such labels. We hereby ask all interested parties around the world who wish to start an environmental labelling program in their country to benefit from our intellectual assistance and support in the form of consulting contracts. Increasing human awareness of the urgent need to protect the environment has led to changes in all levels of activities, including the production of marketing products, consumption, use, and sale of goods and services at the national and international levels. Stakeholders involved in environmental protection include consumers, producers, traders, scientific and technological institutes, national authorities, local and international organizations, environmental gatherings, and human society in general. Decisions by consumers and sellers of products are made not only on the basis of key points such as quality, price, and availability of

products but also on the environmental consequences of products, including the consequences that a product can have before, after and during production. The most important environmental consequences include water, soil, and air pollution along with waste generation, especially hazardous waste. Further consequences include noise, odor, dust, vibration, and heat dissipation as well as energy consumption using water, land, fuel, wood, and other natural resources. There are further effects on certain parts of the ecosystem and the environment. In addition, the environmental consequences not only include the natural use of the products but also abnormal and even emergency or accidental uses. The basis of studies and studies in this field is done through product life cycle evaluation, which generally involves the study and evaluation of environmental aspects and consequences of a category (product, service, etc.) because of the preparation of raw materials for production until they are used or discarded. Sometimes the phrase "review from cradle to grave" is used for such an evaluation. In addition to the above, the environmental consequences that may occur at any stage of the product life cycle, including the preliminary stages and its preparation, production, distribution, operation, and sale, should also be considered when evaluating it. This type of evaluation refers to product life cycle analysis from an environmental point of view,"

which is a useful tool for measuring the degree of environmental health of a product, comparing different products, improving product quality, and confirming the environmental health claims of the product. The environmental health analysis tool for products and services facilitates their placement in domestic or foreign markets, considering that the awareness of consumers and retailers about the environmental consequences of the product has increased, as has the accurate and explicit measurement by the people in charge at all levels. Local, national, and international in the field of environmental protection. Products that can claim to be environmentally complete in all stages of their life cycle and meet the mandatory and optional environmental needs are considered successful products. Environmental messages refer to the policies, goals, and skills of product manufacturing companies as part of the environmental management systems in which they are applied, and consumers and retailers are increasingly paying attention to this issue when making purchasing decisions. In addition, companies have been encouraged and even forced to adapt their environmental management systems to agencies and retailers and to local, national, international, and other environmental issues.

The environmental health message of a product can be conveyed to the consumer in various ways, including implicitly or explicitly. For example, the implicit or implicit message conveyed directly by the product to the customer is that the product is suitable for the intended use and purpose, and, without material waste in size, weight, and dimensions, is perfectly proportioned and without additional packaging. Sometimes it is necessary to convey these messages and claims about the correctness of the product quite clearly through magazines or other media as well as through certificates that are accurate, simple, and convincing to the consumer in the form of a label. These messages must be accurate and fact-based; otherwise they will nullify the product and create contradictory effects. Confirmation of these claims by a third-party organization will increase its credibility. It should also be noted that the multiplicity of these messages, depending on the type of products or companies producing them, confuses consumers in the market and also creates artificial boundaries or causes a differentiated distinction against certain products or companies. Various models, principles, and methods have been provided by local, regional, national, and international organizations to demonstrate product life cycle analysis and other guidelines on environmental management systems and their labels. At the national level, significant advances have been made in the design of environmental labels in various countries, including developing countries and the Scandinavian countries. For example, the first project was designated in Germany as a Blue Angel in 1977, later on Canada in 1988, the Scandinavian countries and Japan in 1989, the United States and New Zealand in 1990, India, Austria, and Australia in 1991, And in 1992, Singapore, the Republic of Korea, and the Netherlands developed their national environmental labelling. Environmental labels are an environmental management tool that is the subject of a series of ISO 14000 standards. These environmental labels provide information about a product or commodity in terms of its broad environmental characteristics, whether it is about a specific environmental issue or about other characteristics and topics.Interested and pro-environmental buyers can use this information when choosing products or goods. Product makers with these environmental labels hope to influence people's purchasing decisions. If these environmental labels have this effect, the share of the product in question can increase, and other suppliers may create healthy environmental competition by improving the environmental aspects of their products and commodities. The overall goal of environmental labels is to convey acceptable and accurate information that is in no way misleading regarding the environmental aspects of products and commodities, and they encourage the consumer to buy and produce products that reduce stress on the environment. Environmental labelling must follow the general principles that the International Organization for Standardization has published in a collection entitled the ISO 14020 standard, which refers to these general principles here. It should be noted that other documents and laws in this field are considered if they are in accordance with the principles set out in ISO 14020.

How can we make stationery eco friendly? Making your supplies & stationery stockpile more sustainable

- Make sure your paper has been sourced sustainably.
- Print double-sided and black & white.
- Cork noticeboards.
- Swap pens for pencils where possible.
- Reuse and recycle boxes and packaging.

CHAPTER 2

General Principles on Environmental Labelling

1 The First Principle: Evironmental notices and labels must be accurate, verifiable, relevant, and in no way misleading and/or deceptive.

2 The Second Principle: Procedures and requirements for environmental labels will not be ready for selection unless they are implemented by affecting or eliminating unnecessary barriers to international trade.

3 The Third Principle: Environmental notices and labels will be based on scientific analysis that is sufficiently broad and comprehensive, and to support this claim, the product must be reliable and reproducible.

4 The Fourth Principle: The process, methodology, and any criteria required to support the announcements on environmental labels will be available upon request all interested groups.

5 The Fifth Principle: Development and improvement of environmental notices and labels should be considered in all aspects related to the service life of the product.

6 The Sixth Principle: Announcements on environmental labels will not prevent initiative and innovation but will be important in maintaining environmental implementation.

7 The Seventh Principle: Any enforcement request or information requirement related to environmental notices and labels should be limited to the necessary information to establish compliance with an acceptable standard and based on the notification standards and environmental labels.

8 The Eighth Principle: The process of improving the announcement and environmental labels should be done by an open solution with interested groups. Reasonable impressions must be made to reach a consensus through this process.

9 The Ninth Principle: Information on the environmental aspects of the product and goods related to an advertisement and environmental label will be prepared for buyers and interested buyers from a group consisting of an advertisement and an environmental label.

What is eco-friendly paper?

There is a wide variety of alternative 'fibres' that can work as an alternative to wood-pulp paper. Sources for tree-free paper include: agricultural residues for example, sugar cane bagasse, husks and straw. fibre crops and wild plants – such as bamboo, kenaf, hemp, jute, and flax. textiles and cordage wastes.

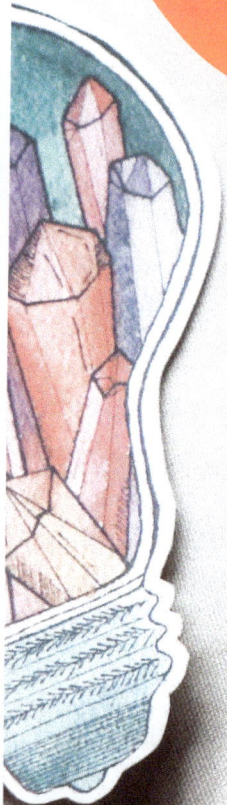

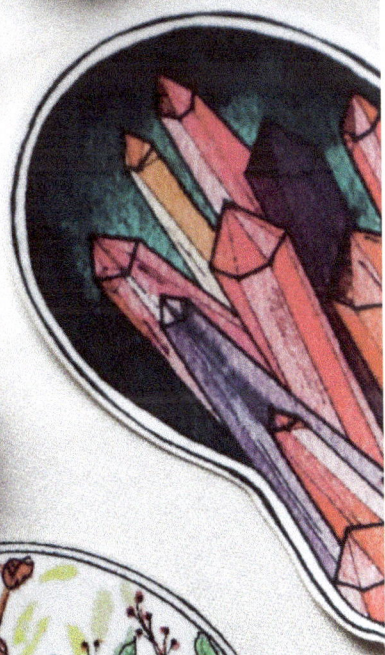

CHAPTER 3

Types of Environmental Labelling

At present, according to the classification provided by the International Organization for Standardization, there are three types of environmental labelling
patterns:

1 Type I labelling: This labelling is known as eco-labelling, and because it
is difficult to translate this word into many languages, it presents another
reason to adhere to a numerical classification system. In the content of
Type I labelling, a set of social commitments that creates criteria according to
the scientific principles on the basis of which a product is environmentally preferable is discussed. Consumers are then instructed in assessing environmental
claims and must decide which packaging is more important.

2 Type II labelling: refers to the claims made on product labels in connection
with business centers. This includes familiar claims such as recyclable,
ozone-friendly, 60% phosphate-free, and the like. This type of labelling
can be in the form of a mark or sentence on the product packaging. Some of them
are valid environmental claims—and some can be completely misleading. Usually, all countries have laws against deceptive advertisements, so why has the
International Organization for Standardization discussed this issue? The answer
is that it is not clear whether the environmental claims have a technical basis or
whether the ad is meaningless.

3 Type III labelling: is a distinct form of third-party environmental labelling pattern designed to avoid the difficulties that can result from type-one labelling. Technical committee for Environment of International organization for Standardization has undertaken a new project to standardize guidelines and Type III labelling methods. One of the main objections raised by industries to Type I labelling is the basis for its management.

What is eco friendly notebook?

Eco friendly notebooks use FSC certified paper, and are entirely made from degradable and recyclable materials (the covers aren't recyclable, but will degrade in a landfill). They're also 100% vegan!

CHAPTER 4

Type I Environmental Labelling

Type I labelling: This labelling is known as eco-labelling, and because it is difficult to translate this word into many languages, it presents another reason to adhere to a numerical classification system. In the content of Type I labelling, a set of social commitments that creates criteria according to the scientific principles on the basis of which a product is environmentally preferable is discussed. Consumers are then instructed in assessing environmental claims and must decide which packaging is more important.

Type I adhesive has the following specifications:
A. Has an optional third-party template.
B. When the product meets a certain standard, the labelling of this product is included.
C. The purpose of this program is to identify and promote products that play a pioneering role in terms of environment, which means its criteria are at a higher level than the average environmental performance.
D. Acceptance/rejection criteria are determined for each group of products and are publicly available.
E. The criteria are adjusted after considering the environmental consequences of the product life cycle.

Examples of Type I Labelling:
In this section, and considering the importance of this type of labelling, I provide a description of some examples of Type I labelling related to some countries along with a   list of products on which this mark is placed.

Germany

FSC® is a global not-for-profit organization that sets the standards for responsibly managed forests, both environmentally and socially. When timber leaves an FSC certified forest they ensure companies along the supply chain meet our best practice standards also, so that when a product bears the FSC logo, you can be sure it's been made from responsible sources. In this way, FSC certification helps forests remain thriving environments for generations to come, by helping you make ethical and responsible choices at your local supermarket, bookstore, furniture retailer, and beyond. www.fsc.org

FSC® International
Adenauerallee 134
53113 Bonn
E-mail: info@fsc.org
Phone: +49 (0) 228 367 66

FSC Canada
50 rue Sainte-Catherine Ouest,
bureau 380B, Montreal, QC H2X 3V4
Email: info@ca.fsc.org
Telephone: 514-394-1137

Ukraine

The ecolabelling program in Ukraine was founded on the initiative of the All-Ukrainian NGO "Living Planet" in 2003. The Green Crane is the first and the only one Type 1 Ecolabel in Ukraine that recognized officially.

The main objective of company's activity is to evaluate the products for compliance with environmental criteria according to ISO 14024 scheme in order to ensure the reliability of data on the environmental benefits of products within a specific category based on the results of the life cycle assessment. Over the 16 years of the program's existence, the Green Crane ecolabel has become a recognizable reliable reference point for consumers and government organizations (in "green procurement" process), as well as effective marketing tool for business.

Program Statistical Information. Today, the program operates with 57 certification standarts in various industries - construction, food, chemical, textile and other. More than 500 certificates have been issued throughout the program's history. Currently, 68 certificates for more than 1,000 products are valid.

Contact:
NGO «Living Planet»
Email: os@ecolabel.org.ua,
info@ecolabel.org.ua
Tel: +380 44 332 84 08
Adress: Magnitogorsky Lane1-B, Kyiv, Ukraine - 02094
Web: https://www.ecolabel.org.ua/en

Global

As the global safety science leader, UL provides the expertise, insights and services necessary to solve critical safety and business challenges. We help our customers achieve their safety, security and sustainability goals, meet quality and performance expectations, manage risk and achieve regulatory compliance.

In the course of our work we meet extraordinary people whose companies are having an extraordinary impact on the world and creating the future. UL's rigorous scientific processes, experience and solutions empower our customers to innovate fearlessly and drive positive change. We never stop working for a safer world and our offerings continue to evolve with advancements in science and technology. We provide testing, inspection and certification (TIC), training, advisory and risk management services, decision-making tools and intelligence to help our customers, based in more than 100 countries, meet important business objectives. All of our offerings have one connecting thread: trust. Confidence drives commerce, and trust in innovation is essential to market access, business success and better living. UL empowers trust. One of the many specialty areas that UL focuses on are furniture and building materials, which emit chemicals into the air that people breathe indoors. These chemicals can trigger asthma, headaches and allergy attacks, which is why UL established its GREENGUARD Certification program. Products that have achieved UL's GREENGUARD Certification are scientifically proven to meet some of the world's most rigorous third-party chemical emissions standards, helping to reduce indoor air pollution and the risk of chemical exposure. UL conducts ongoing Certification retesting for more than 10,000 chemicals, not just one time only. Upon seeing a GREENGUARD Certification, consumers can feel confident that the product they purchase is low-emitting. GREENGUARD Gold Certification is the highest/most stringent level of certification. It was developed with the specific health sensitivities of children and other vulnerable populations.

Complete contact detail that readers can contact you : Readers can contact UL anytime through the form on this website:

https://www.ul.com/contact-us

China

China Environmental United Certification Center (CEC), approved by the Ministry of Ecology and Environment of the People's Republic of China (MEE) and accredited by Certification and Accreditation Administration Committee of PRC, is a comprehensive certification and service institution leading in environmental protection, energy saving and low carbon areas. . CEC is committed to serve building national ecological civilization; and has carried out research on environmental protection, energy saving, low carbon development strategies and solutions; has been continuously improving and innovating green industry evaluation system on industrial green development and transition CEC is building a bridge between green production and green consumption by offering independent, impartial and high-quality evaluation and certification service for government, enterprises and the public. CEC is a state-owned, non-profit, legal entity of independent third-party certification. It integrates the certification resource from the former National Accreditation Center for Environmental Conformity Assessment, the Secretariat of China Environmental Labelling Products Certification Committee, Environmental Development Center of MEE, the Chinese Research Academy of Environmental Sciences and other institutions. Business areas includes: products certification, management systems certification, services certification, addressing climate change, energy-saving and energy efficiency certification, green supply chain assessment, environmental stewardship, green credit assessment and green manufacturing system evaluation. CEC also carries out standard establishment and research project and international cooperation and exchanges, etc.

Contact:
Website: http://en.mepcec.com/
E-mail: zhangxiaoh@mepcec.com , zhangxiaoh@mepcec.com

Sri Lanka

National Cleaner Production Centre (NCPC), Sri Lanka was set up by UNIDO in 2002, as a project under the Ministry of Industry to provide the technical expertise and support to the industry and business enterprises in order to prevent pollution and conserve resources by the application of Cleaner Production (CP) and other proactive environmental management tools. NCPC Sri Lanka is registered as a Company by Guarantee not for profit organization under the Act No. 7 of 2007. Over the past two decades, it has evolved as the foremost sustainability solution provider in the country.

The ISO 9001:2015 certified Centre is a registered Energy Service Company (ESCO) under Sustainable Energy Authority (SEA) and a registered consultant under Central Environmental Authority (CEA). It is a founding member of UNIDO/UNEP Resource Efficient and Cleaner Production Network (RECP Net), a global family of 52 NCPCs. NCPC Sri Lanka is a member of Climate Technology Centre & Network (CTCN) and associate member of Global Eco-labelling Network (GEN). Accordingly, we at National Cleaner Production Centre (NCPC), Sri Lanka has developed Eco Labelling scheme under the ISO 14024:2018 - Environmental labels and declarations. NCPC Eco labelling scheme developed, with the Support of United Nations Environment Programme, Under One Planet Network Consumer Information Programme for Sustainable Consumption and Production (CI-SCP).

Contact:
Tel: +94 11 2822272/3,
Fax: +94 11 2822274
E mail: info@ncpcsrilanka.org
Web: www.ncpcsrilanka.org

Hong Kong

The Green Council is a non-profit, tax-exempt charitable environmental stewardship organisation and certification body (Reg. No.: HKCAS-027) of Hong Kong established in 2000. A group of individuals from different sectors of industry and academics shared the vision to help build Hong Kong into a world-class green city for the future. They formed the Green Council with the aim of encouraging the commercial and industrial sectors to include environmental protection in their management and production processes. The Green Council is a non-profit, tax-exempt charitable environmental stewardship organisation and certification body (Reg. No.: HKCAS-027) of Hong Kong established in 2000. A group of individuals from different sectors of industry and academics shared the vision to help build Hong Kong into a world-class green city for the future. They formed the Green Council with the aim of encouraging the commercial and industrial sectors to include environmental protection in their management and production processes. The Green Council is a non-profit, tax-exempt charitable environmental stewardship organisation and certification body (Reg. No.: HKCAS-027) of Hong Kong established in 2000. A group of individuals from different sectors of industry and academics shared the vision to help build Hong Kong into a world-class green city for the future. They formed the Green Council with the aim of encouraging the commercial and industrial sectors to include environmental protection in their management and production processes.

Contact:
Website: https://www.greencouncil.org/hkgls
Email: info@greencouncil.org
Telephone: (852) 2810 1122

BIO LATINA
CERTIFICADORA

Peru

BIO LATINA, the consolidated byproduct of four Latin American national certification entities.Since 1998, we have provided certification services in Latin America for national and international markets. We seek to help create a more sustainable and resilient world. With these goals in mind, we have expanded our service portfolio beyond organic to social and environmental certifications.

Visit us: https://biolatina.com

From our regional offices we serve Latin American.

Our headquaters:
Av. Javier Prado Oeste 2501, Bloom Tower Of. 802, Magdalena del Mar,
 Lima 17, Perú

EKOagros

ORGANIC CERTIFICATION

Lithuania

EKOAGROS is the only institution in Lithuania for more than 20 years carrying out certification and control activities of organic production and products of national quality, also providing services of certification activities in accordance with the foreign national and private standards in foreign countries. From year 2017 EKOAGROS is accredited as certifying agent to conduct certification activities on crops, wild crops, livestock and handling operations in accordance with USDA NOP.

Contact information:
EKOAGROS
Address K. Donelaicio str. 33, LT-44240 Kaunas, Lithuania
Tel. No. +370 37 20 31 81
Website: www.ekoagros.lt

Korea
Eco-Label

Republic of Korea

The Korea Eco-labelling is a certification system enforced by the Ministry of Environment and KEITI(Korea Environmental Industry & Technology Institute). Since its foundation in April 1992, the system has certified a wide range of eco-friendly products, which were selected as excellent not only in terms of their environmental-friendliness, but also for their quality and performance during their life cycle. Korea Eco-labelling is voluntary certification scheme to attach logo to products with superior environmental quality throughout their lifecycle to other products of the same use, and thus to provide product information to consumers. For 30 years, the scheme has launched plenty of eco-labelling product standards covering personal and household goods, construction materials, office equipment furniture, etc. It products categories which cover all aspects of products, such as reduction of use of harmful substances, energy saving, resource saving, etc. As of April 30th 2021, 169 criterias(=standards), and certifications for 18,250 products(4,549 companies) have maintained.

Contact:
Korea Environmental Industry & Technology Institute(KEITI)
Office of Korea Eco-Label Innovation
Address: 215, Jinheung-ro, Eunpyeong-gu, Seoul, Repulic of Korea
T: +82 2 2284 1518
F: +82 2 2284 1526
E: accolly@keiti.re.kr
W: www.keiti.re.kr

USA

The Carbonfree® Product Certification is a meaningful, transparent way for you to provide environmentally-responsible, carbon neutral products to your customers. By determining a product's carbon footprint, reducing it where possible and offsetting remaining emissions through our third-party validated carbon reduction projects, companies can:

- Differentiate their brand and product
- Increase sales and market share
- Improve customer loyalty
- Strengthen corporate social responsibility & environmental goals

The Carbonfree® Product Certification Program is proud to be part of Amazon's Climate Pledge Friendly Program!
Carbonfund.org is leading the fight against climate change, making it easy and affordable to reduce & offset climate impact and hasten the transition to a clean energy future.

Contact:

O: 240.247.0630 ext 633
C: 203.257.7808
M: 853 Main Street, East Aurora, NY, 14052

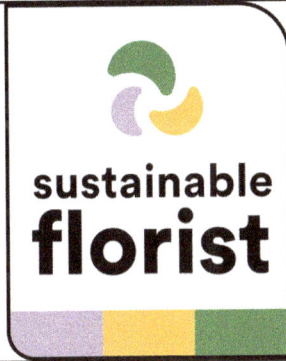

sustainable florist

Netherland

For more than 25 years, the independent Dutch foundation SMK works from professional knowledge with companies to improve the sustainability of products and business management. SMK cooperates with an extensive stakeholder network of governments, producers, branch and non-governmental organisations, retailers, consultancies, researchers. The SMK Boards of Experts establish objective criteria for more sustainable products and services. SMK's transparent work processes, third party audits and certifications are conducted according to international certification standards, mostly under supervision of the Dutch Accreditation Council. Besides, SMK is Competent Body of the EU Ecolabel. SMK keeps an extensive database of sustainability criteria.

Contact:
Bezuidenhoutseweg 105 - 2594 AC Den Haag
Telefoon: 070-3586300
Mobiel: 06-82311031
(niet op woensdag)
www.smk.nl

Taiwan

The Green Mark GM) Program was launched by the Environmental Protection Administration of Taiwan (TEPA) in 1992. As the official Type I eco-labeling program, it is in compliance with the requirements of the international stadard, ISO 14024 and is considered an important tool to promote green consumption and production .

To improve the GM application/review mechanism and introduce a third party certification scheme, TEPA promulgated the «Guideline for the Management of Certification Organizations for Environmental Protection Products" in June 2012. Both Environment and Development Foundation (EDF) and the Taiwan Testing and Certification Center (ETC) were commissioned by TEPA as official certifiers. With the expansion of certification capacity and authorization of the certification decision, the certification time was greatly reduced.

Contact :

Website: www.edf.org.tw
TEL: 886-3-5910008 #39
E-mail: lhliu@edf.org.tw

Denmark, Finland, Norway, Iceland, Sweden

The Nordic Swan Ecolabel

The Nordic Swan Ecolabel is the official Nordic ecolabel supported by all Nordic Governments. It is among the world›s strictest and most recognised environmental certifications.

The Nordic Swan Ecolabel is a Type I environmental labelling program established in 1989 by the Nordic Council of Ministers, connect¬ing policy, people, and businesses with the mission to make it easy to make the environmentally best choice. Nordic Ecolabelling is the non-profit organisation responsible for the Nordic Swan Ecolabel.

The organisation offers independent third-party certification and support for a wide range of product areas and services, ensuring that they comply with the Nordic Swan Ecolabel's strict requirements through documentation and inspections.

30 years of experience and expertise has made the Nordic Swan Ecolabel a powerful tool that paves the way to a sustainable future by giving producers a recipe on how to develop more environmentally sustainable products, and giving consumers credible guidance by helping them identify products that are among the environmentally best.

Globally, you can find more than 25,000 Nordic Swan ecolabelled products. 93% of all Nordic consumers recognise the Nordic Swan Ecolabel as a brand, and 74% believe that the Nordic Swan Ecolabel makes it easier for them to make envi¬ronmentally friendly choices (IPSOS 2019).

Denmark, Finland, Norway, Iceland, Sweden

Securing a sustainable future

The Nordic Swan Ecolabel works to reduce the overall environmental impact from production and consumption and contributes significantly to UN Sustainable Development Goal 12: Responsible consumption and production.

To ensure maximum environmental impact, the Nordic Swan Ecolabel sets product specific requirements and evaluates the environmental impact of a product in all relevant stages of a product lifecycle - from raw materials, production, and use, to waste, re-use and recycling.

Common to all products certified with the Nordic Swan Ecolabel is that they meet strict environmental and health requirements. All requirements must be documented and are verified by Nordic Ecolabelling. Nordic Ecolabelling regularly reviews and tightens the requirements.

Therefore, certifications are time-limited and companies must re-apply to ensure sustainable development.

International website:
Nordic-ecolabel.org
National websites:
Denmark: ecolabel.dk
Sweden: svanen.se
Norway: svanemerket.no (in Norwegian)
Finland: joutsenmerkki.fi (in Finnish)
Iceland: svanurinn.is (in Icelandic)

Thailand

The Thai Green Label Scheme was initiated by the Thailand Business Council for Sustainable Development (TBCSD) in October 1993. It was formally launched in August ١٩٩٤ by The Thailand Environment Institute (TEI) and Thai Industrial Standards Institute (TISI). The Green Label is an environmental certification logo awarded to specific products which have less detrimental impact on the environment in comparison with other products serving the same function. The Thai Green Label Scheme applies to all products and services, but not foods, beverage, and pharmaceuticals. Products or services which meet the Thai Green Label criteria may carry the Thai Green Label. Participation in the scheme is voluntary.

Thailand Environment Institute (TEI)
16/151 Muang Thong Thani, Bond Street,
Bangpood, Pakkred, Nonthaburi 11120 THAILAND
Tel. +66 2 503 3333 ext. 303, 315, 116
Fax. +66 2 504 4826-8
Website: http://www.tei.or.th/greenlabel/
Email: lunchakorn@tei.or.th

EUROPE

Established in 1992 and recognized across Europe and worldwide, the EU Eco-label is a label of environmental excellence that is awarded to products and services meeting high environmental standards throughout their life-cycle: from raw material extraction, to production, distribution and disposal. The EU Eco-label promotes the circular economy by encouraging producers to generate less waste and CO_2 during the manufacturing process. The EU Ecolabel criteria also encourages companies to develop products that are durable, easy to repair and recycle.

The EU Ecolabel criteria provide exigent guidelines for companies looking to lower their environmental impact and guarantee the efficiency of their environmental actions through third party controls. Furthermore, many companies turn to the EU Ecolabel criteria for guidance on eco-friendly best practices when developing their product lines. The EU Ecolabel helps you identify products and services that have a reduced environmental impact throughout their life cycle, from the extraction of raw material through to production, use and disposal. Recognised throughout Europe, EU Ecolabel is a voluntary label promoting environmental excellence which can be trusted.

Spain , Germany, Italy, Sweden, Greece, Portugal, Poland, Belgium, Netherlands, Estonia, Finland, Austria, Lithuania, Czech Republic, Norway, Cyprus, Ireland, Slovenia, Hungary, Romania, Croatia, Bulgaria, Malta, Slovak Republic, Latvia, Luxembourg, Iceland

Contact and more information via: http://ec.europe.eu

There is no better time than now to modify your daily activities and make them more eco-friendly. Choosing eco-friendly office supplies and providing sustainable stationery products to your kids can be a great place to begin. You may not be able to switch your entire office or school supplies at once but you can always choose environment-friendly products when you have to.

Eco-friendly office stationery

To start a sustainable office environment, simply make a list of all the supplies you use and research earth-friendly alternatives. The following are some of the best environmentally friendly supplies available to help reduce your office's carbon footprint.

Recycled paper products

Choose recycled paper products or products that can be recycled after use. From envelopes to compostable paper cups and plates, recycled paper products now come in many forms. A sustainable change in this department will have a significant positive impact considering how frequently paper products are used in the office.

Compostable trash bags

Using non-biodegradable trash bags week after week is not good for the environment as it all end up as waste. Trash bags made of post-consumer recycled (PCR) material is much better for the environment. Other biodegradable trash bags are also available. To make full use of it, make sure the trash bags are filled up to the fullest before you replace it.

Eco-friendly pencils and pens

Using biodegradable pens made from renewable or recycled material is a great way to make your office eco-friendly. You can also choose refillable pens to reduce some waste or use pencils made from 100 percent recycled newspaper. You can even print your company logo on eco-friendly pens to give potential customers a great first impression.

Non-toxic cleaner

Why introduce formaldehyde and other harmful chemicals in your confined office floor? There are plenty of environmentally friendly cleaning options available to clean all of the surfaces in your office while keeping the air quality clean. (for more detail refer to IEL Vol.5)

Solar USB chargers

Say goodbye to wall sockets and use solar power to charge your phone. These solar-powered USB chargers can help to save energy and also personalizes your phone charging station. This is an inexpensive transition for people who leave their phones plugged in all the time. (for more detail refer to IEL Vol.2)

Eco-friendly school stationary

Children learn a lot from their surroundings and day-to-day work. So, why not educate them in an eco-friendly setting. The goal is to provide your child with stationery supplies that are recycled or environmentally friendly.

CHAPTER 5

Type II Environmental Labelling

Type II environmental labelling refers to the claims made on product labels in connection with business centers. This includes familiar claims such as recyclable, ozone-free, 60% phosphate-free, and the like. This type of labelling can be in the form of a mark or sentence on the product packaging. Some of them are valid environmental claims—and some can be completely misleading.

Usually, all countries have laws against deceptive advertisements, so why has the International Organization for Standardization discussed this issue? The answer is that it is not clear whether the environmental claims have a technical basis or whether the ad is meaningless.

Most countries have guidelines at the national level to help producers and consumers know what constitutes a true, scientifically valid claim.
There is a national standard on this in Canada. In Australia, the Consumer Commission has published guidance on this, and there are similar examples in other countries.

Canada

Environmental Sustain for Future kids established in Vancouver, BC Canada in 2020. (ESFK) is an international ecolabel focused on taking care of environment for future of kids.

ESFK defined as 'self-declared' environmental claims made by manufacturers and businesses based on ISO 14020 series of standards, the claimant can declare the environmental objectives and targets in relation to taking care of environment for future kids. However, this declaration will be verifiable.

Environmental Sustain for Future Kids
Vancouver, BC CANADA

Email: info@esfk.org
Web: www.esfk.org

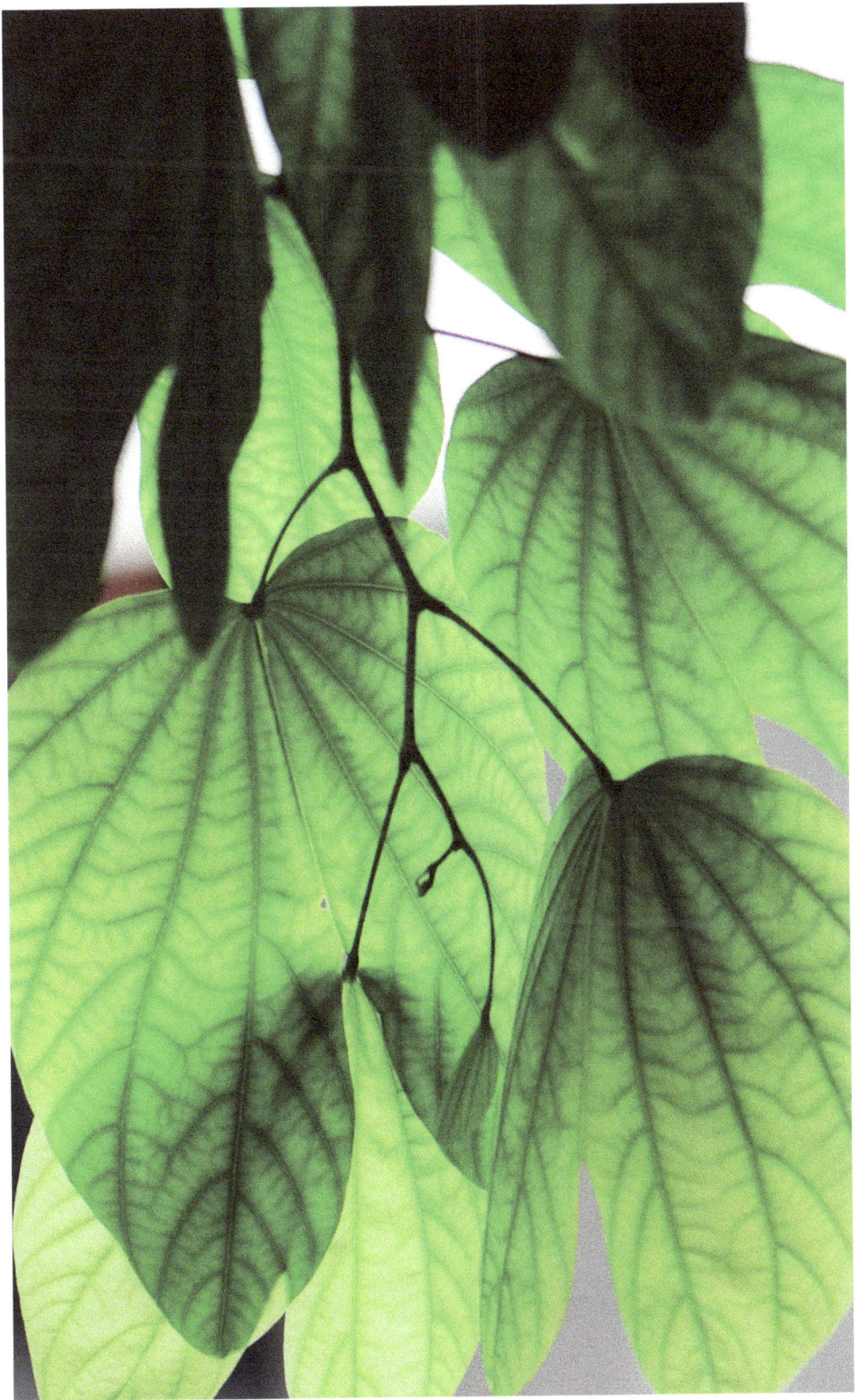

**Please provide your child, emplo
envi**

h stationery supplies that are recycled or
ntally friendly.

To:

All Schools, Libraies,
offices, companies, & ....
all over the globe

CHAPTER **6**

Type III Environmental Labelling

Type III environmental labelling is a distinct form of third-party environmental labelling pattern designed to avoid the difficulties that can result from type I labelling. Technical committee for Environment of International organization for Standardization has undertaken a new project to standardize guidelines and Type III labelling methods. One of the main objections raised by industries to Type I labelling is the basis for its management.

Due to the nature of the system, less than 50% of the various products on the market can meet the criteria and qualify for Type I Labelling. As long as the industry is the main supporter of other third-party models for quality systems, it is sometimes difficult for an industry to support a program that can only benefit 15% of its members. This type of labelling is currently practiced in some countries, such as Sweden, Canada, and the United States. Choosing the right product has never been easy, but Type III labelling will help because each product can have a label that describes its environmental performance and is certified by a third-party company. Consumers can then compare labels and choose their favorite products.

CHAPTER 7

All about 'Eco-friendly' Wood & Stationery Products

Eco-friendly stationery is a broad category that encompasses recycled and sustainable stationery, along with zero-waste gifts like eco-friendly notebooks that can be fully recycled once they have been used. The term applies to all kinds of green stationery, from eco-friendly pens and pencils, to recycled notebooks and ethical stationery gifts – and different products might have different ethical and environmental credentials. For example, some eco stationery may be made from ethically produced materials like sustainably managed timber, while others might use more recycled raw materials or be designed in such a way to avoid ending up in a landfill site.In this volume, we'll look in more detail at what makes green stationery 'green' and why it makes sense to buy eco-friendly stationery for yourself and others.

What is sustainable stationery?

There's no strict definition of sustainable stationery. It could be recycled or recyclable (or both), it might use raw materials that would otherwise go to landfill, or it might be ethical in some other way. Because it's such a broad definition, there's plenty of choice, allowing you to choose a sustainable pen that fits your lifestyle and your own eco priorities.

Why should I try sustainable stationery?

There's really nothing to lose. Sustainable stationery often does not cost any more than its traditional equivalents, and there are some really stunning designs of eco pens, pencils and notebooks to choose from.

YOU CAN MAKE A CONSCIOUS EFFORT TO BE ENVIRONMENTALLY FRIENDLY

If you want to take action for sustainable living, without making major changes to your lifestyle, eco stationery is a simple first step to take. Sustainable stationery products function exactly as normal and there's usually no visible difference either, so you get peace of mind without compromise.

IT SETS A GOOD EXAMPLE

As well as improving your personal eco-profile, you can encourage others to take their own action for sustainable living too. International Environmental Labelling Book series of 11 volumes is a great gift for this scenario; by giving it to friends, family and co-workers. You can give this book series (look at the end of this book for more detail), an eco-friendly notebook, pen or pencil and send them on their own path towards a more eco-friendly life-style.

YOU WILL REDUCE WASTE

Sustainable stationery is typically designed to last longer. In fact, any refillable pen will generate less waste than a disposable pen, so opt for a cartridge pen, traditional fountain pen or refillable ball pen if you want to do your bit.

If you are in business, you could consider giving a customised eco-friendly gift, to encourage your customers to cut down on their use of disposable stationery too.

How does eco-friendly stationery help the environment?

Eco-friendly stationery (depending on the product) can divert materials away from landfill during its manufacture, and reduce landfill waste due to disposable plastic pens being thrown away. Sustainable stationery is also more likely to use renewable raw materials such as wood and eco plastics, rather than single-use plastics made from fossil fuels.

How is sustainable stationery made?

Green stationery is made from sustainable raw materials – and these, in turn, are produced using ethical, long-term sustainable practices. For example, One of the the largest pencil brand in the world, has a pioneering forestation project in Brazil and Colombia where two million trees are planted every year. Whenever a tree is felled to provide timber for more than the 2.2 billion pencils manufactured each year, a new one is planted. This is just one example of how sustainable stationery can be made and managed.

Which sustainable stationery products should I try?

Ecolabel programs authorize the use of environmental logos on products or services that meet a strict set of criteria. These ecolabels indicate an overall environmental preferability of a product or service within a particular product or service category based on life cycle "considerations," although not necessarily a more complex full life cycle assessment. Some ecolabels are created and managed on a national level while others are international in scope. They may be administered by government bodies or private sector labelling standards organizations, and typically involve certification by legitimate and independent third party organizations.

Final thoughts

We could all live a little more sustainably, but simple steps like using more sustainable stationery allow us to do so without compromising on our existing lifestyles in any significant way. Many people have already swapped plastic drinking straws for metal, bamboo or paper equivalents. Switching to green stationery is an obvious next step. And with such a great selection of eco pens and pencils, eco-friendly notebooks and zero-waste gifts available, there's no reason to delay investing in a great-quality pen or pencil that will stay by your side for many years to come.

19 Eco Friendly And Zero Waste School Supplies

New stationery and school supplies has always been an exciting time for most kids (and many adults too!). Writing this article brings back memories of fresh stationery and the joys of organizing and reorganizing pens, pencils, rulers, notebooks…Although, for parents and adults who are aiming to keep school sustainable though, it can bring a sense of dread. Plastic packaging, toxic components, and boxes of unused junk. To help, here are a few strategies for sustainable, low or zero waste school supplies and stationery:

Use what you already have. Scrounge up loose pens and forgotten notebooks (they're usually full of mostly blank pages!).

For things you know you no longer need, consider donating any school supplies that still have a useful life and recycling those that don't.

See what you can find secondhand. eBay is a good place to look for secondhand online school supplies like books, graphing calculators, and other technology (now that many classrooms and home learning programs require tablets). For anything you weren't able to check off with the first two, opt for a zero waste online store or general ethical online shop for eco-friendly back to school supplies as an opportunity to start teaching your kids about conscious consumerism.

For items you end up having to buy new, this list of the best eco-friendly school supplies (including sustainable stationery) will hopefully help. We've tried to find at least one solid sustainable item to fill each major common item. Most of us need office supplies at some point or another, and the two categories are really all the same. So no matter how old you are, there's just as much reason to make sure your office (and school) supplies are eco-friendly.

1. Plantable Pencils

Sustainable FSC certified wooden pencils are zero waste and have a non-toxic natural clay and graphite core. Then, instead of an eraser, they're capped with a biodegradable seed capsule. When your pencil gets too stubby to write with, just stick it in some soil as per their planting guide. Use the crafting potential of these to teach your child some apartment gardening basics! Choose from plain graphite, colored pencils, and inscribed sets. The Mindful Thoughts edition, which bears phrases like "All of us need to grow continuously in our lives", might even help get those creative juices flowing.

2. Recycled Newspaper Pencils

If you're worried about your kids losing their pencils, another eco-friendly pencil alternative is "tree-free" recycled newspaper pencils. These HB soft graphite pencils are comprised entirely of recycled newspapers and magazines. No wood at all, and they look pretty to boot. With an easy peel-to-sharpen design, these are great for kids. Just make sure to help them compost the peels to be truly zero waste. Sets of 5 or 10 come packaged in compostable, unbleached paperboard boxes.

3. Bamboo Pencil sharpener

Sharpen your eco-friendly pencils with this double hole eco-friendly pencil sharpener made of sustainably sourced bamboo and recycled stainless steel. At the end of its life, remove the blades for recycling and compost the bamboo body. It comes packaged with an unbleached cardboard backing printed with soy inks. The only slight downside is the recycled (and recyclable) plastic bit that holds the sharpener inside. This means it's not totally zero waste, but it was the closest we found on the market.

4. Natural Eraser

The dual-sided eraser made of all-natural rubber latex (to erase pencil) and natural silica sand (to erase ink and some markers). One zero waste eraser for all your mistakes. The individual erasers are packaged in a protective, 100% recycled pulp sleeve, which can be composted. Avoid the multi-packs because those are bound together in plastic packages.

5. Natural Grass Pen

Zero waste pens still leave us drawing a bit of a blank. While more sustainable pens exist now, they're still largely wasteful and greenwashing is still a concern, too. The truly best zero waste pens are refillable fountain pens. Great for the office, but perhaps less so for the classroom. That said, you can pick up a fountain pen for a reasonable price and while not totally zero waste either, we feel it's still one of the best alternatives. A good balance of very low waste while still being affordable. If you look after them they'll last a very long time. And you can sometimes find great secondhand fountain pens to reduce your waste further. Otherwise A Natural Grass Pen is a good option for kids. Just remember to gather up all those pens lying around the house and/or office, and use them first. When you're done so they can be recycled!

6. Organic Cotton Pencil Case

Now you need something to store all those eco-conscious goodies in, If you don't have a ditty bag lying around the house check out this line of eco-friendly school supplies from Canada. Their 100% organic cotton zippered pencil bags are available in tons of colors. Each bag is handwoven by underprivileged women in rural area in different countries, so as to provide jobs to these communities. The thick, ultra durable weave is designed to last, even when it gets buried under books in a backpack.

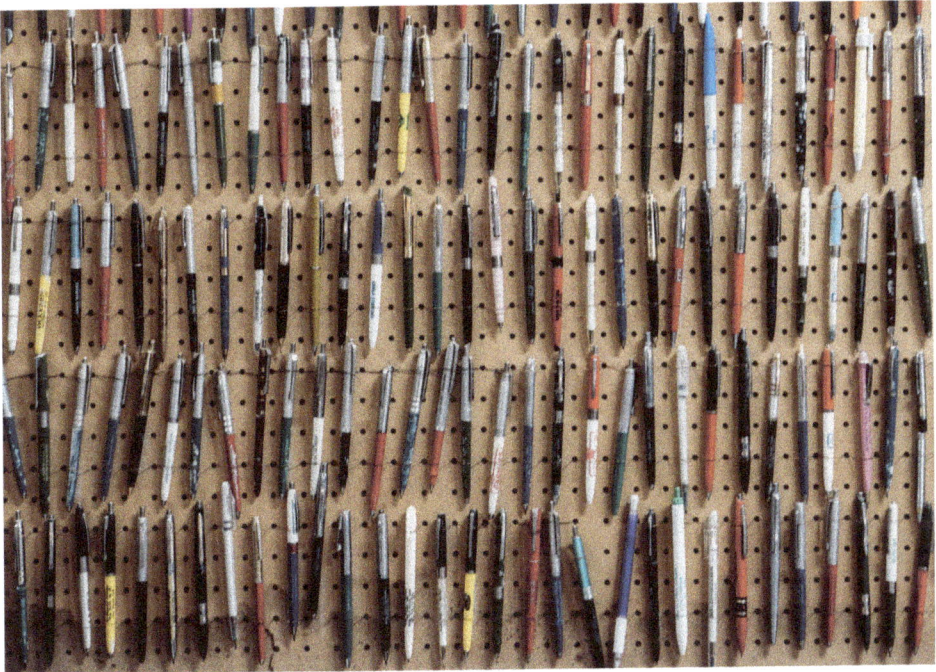

7. Decomposition Books

The best sustainable notebooks we found are different type of Decomposition books. These lovely college-ruled decomposition books are an excellent non-toxic and compostable note-taking solution. Each page is made of 100% post-consumer recycled paper. With so many nature-inspired cover designs (which are printed using soy inks), you can get something different for each subject or mood. Producer also makes spiral-bound notebooks, though we suggest avoiding if possible. Spiral bindings are inherently more wasteful.

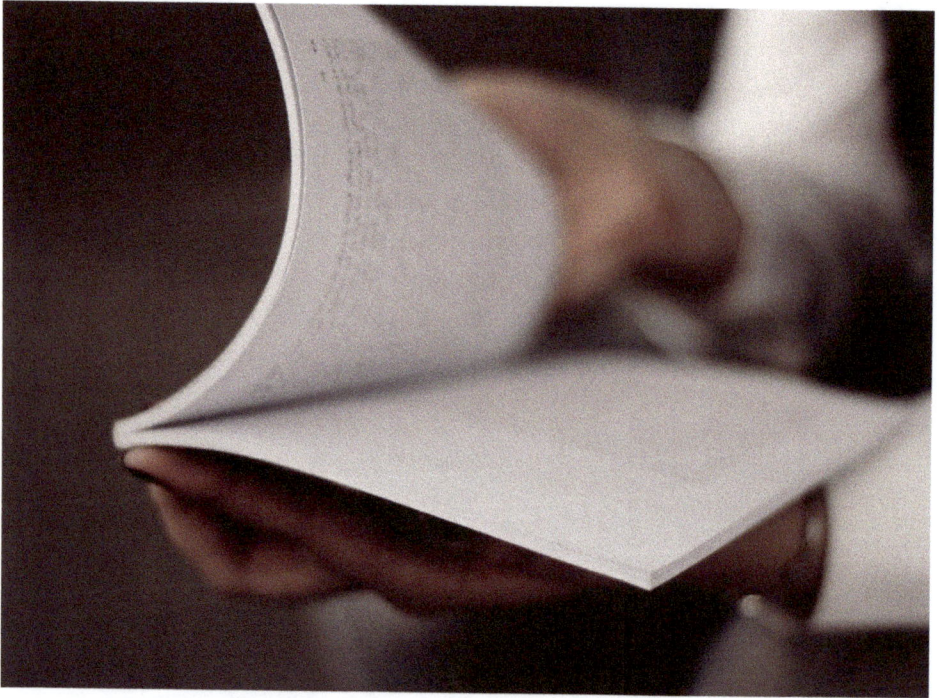

8. Jotter Notebooks

Simple and elegant, these 30% recycled notebooks bound with wax thread are good for both school, work, sketching, journaling, or just making lists. Choose between blank, lined, or 3mm grid dot sheets (which would be perfect as graphic paper for math class). These zero waste notebooks measure 5" x 7" and come with a choice of either 30 or 60 sheets. You can also order them either individually or in packs of three or six. After you're done with the notebook, you can compost the pages and plant the seed paper cover to grow wildflowers! The embedded seeds will grow a blend of Snapdragon, Bird's Eye, Black-Eyed Susan, Clarkia, Sweet Alyssum, and Catchfly.

9. Recycled ReBinder

An undyed zero waste binder is never going to be as cool as the Trapper Keepers you coveted as a kid. But the planet will thank you! No more synthetic fabric-covered or plastic binders (one of many types of plastic, which are lucky to last even a semester). These recycled chipboard ReBinders are a far more durable choice, with a better end of life when they do wear out. Easily screw out the rings for recycling and throw the rest in the compost! And you could even draw on your own designs!

With 0.5" rings, you can fit about 100 sheets of recycled loose leaf paper. Fortunately, that's now pretty easy to find at just about any major office supply store.

10. 2-Pocket Folder

Unbleached folders are made using 22-point chipboard, which is 100% recycled FSC-certified post-consumer waste. This material is thick and durable, eliminating the need for any toxic coatings, including acids that can harm papers. Plus, then your kid (or you for that matter) can decorate it as they like. They're sold in sets of 25 to minimize shipping and packaging.

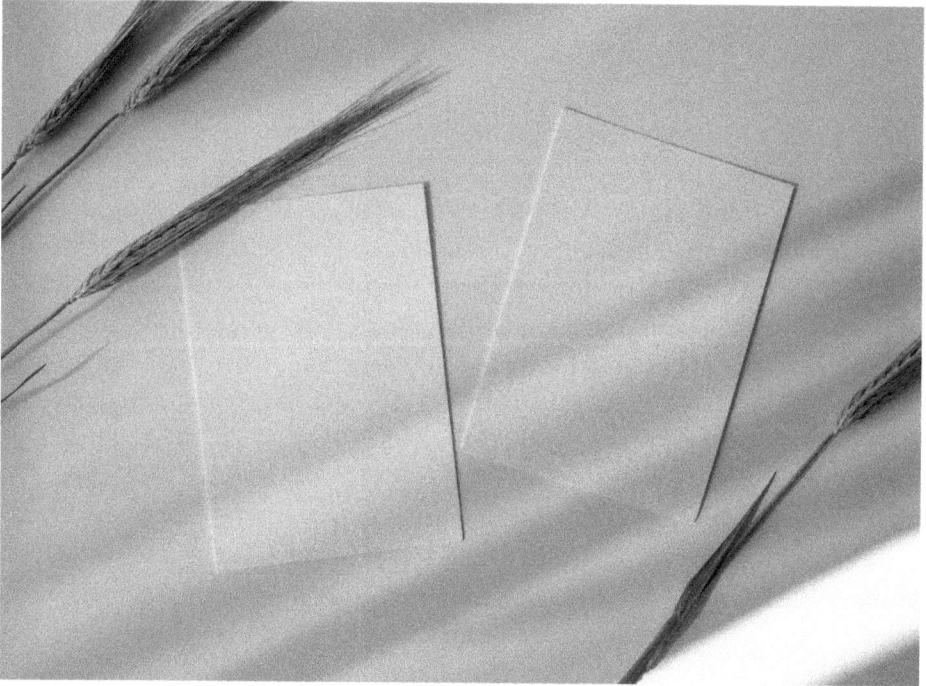

11. Recycled Tab Dividers

About Package Free Shop Recycled and Zero Waste Tab Dividers Your zero waste binder may help keep the planet clean, but how do you keep it clean and organized? With 3-ring compostable chipboard tab dividers, of course! These 5 or 8 tab sets are made out of 85% post-consumer and 15% post-industrial recycled content. With a 13-point thickness, they're durable and should last well beyond one school year. When they do wear out, you can home compost them.

12. Recycled Copy Paper

Not only is Printworks paper made of 100% recycled post-consumer waste (specifically from food and beverage containers), it's FSC-certified and chlorine-free. Unlike many recycled paper products that get shipped to Asia and back again, the waste is collected and remade entirely in the USA (only 300 miles from the mill, in fact). This reduces tons of carbon emissions from shipping. And since it comes in 20-pound boxes, you won't be needing to reorder anytime soon, either. It's just as affordable as traditional copy paper, too. We shouldn't have to choose between paying ourselves and making the planet pay.

13. Staple-Free Stapler
We don't have to prick our fingers pulling out staples before composting or recycling anymore? The PLUS Paper Clinch uses a unique folding technique to bind up to 10 sheets of paper just as staples would… but without all the waste and complication. It's also portable, kid-safe, and, according to the reviews, easy to use due to the ergonomic design and minimal force required. Unfortunately, the body is plastic,but if you take care to make it last, you're at least saving the staples and making whatever you're stapling easier to compost!

14. Paper Tape Cuts
Lasting Things paper tape cuts are made from upcycled 1970s vintage military surplus kraft paper combined with all-natural, compostable acacia gum. To activate the adhesive, just moisten a little bit and stick where desired. While this brown tape may not have all the advantages of clear tape, the biodegradable cello tapes can't be home composted. Besides, the kraft paper look gives it all sorts of crafty potential especially if used in combination with zero waste gift wrapping. An order includes 97 cuts of 9" x 2.5" sizes.

15. Eco-Friendly Backpacks

What good are eco friendly school supplies without a way to tote them? When it comes to something that takes quite a beating like a backpack, true quality is what you want, even more than sustainable materials. United By Blue offers tons of affordable eco friendly backpacks that are both. Their also are made of natural materials like organic waxed cotton canvas, or vegetable dyed repreve recycled polyester. With double stitching and DWR finishes, these are designed to last. They have many designs and sizes, many with internal laptop sleeves to function as two items in one.

16. Cork Laptop Sleeve

This minimalist, zero waste and vegan laptop sleeve is a great way to protect your hardware in your bag or on your commute. The cork provides some padding and water resistance while the cotton liner protects from scratches. But what is cork fabric? It happens to be one of the world's most sustainable materials because it's harvested by shaving the bark of cork trees, as opposed to cutting them down. This process can be repeated every nine years when the bark fully regrows, for up to 300 years. At its eventual end of life, just cut out the metal snaps and compost everything else. The many different sizes available ensure you can find something for every shape and size of laptop.

17. Newspaper Colored Pencils

School isn't all reading, writing, and arithmetic. Sometimes it's fun and we want sustainable art supplies to keep that fun, clean and healthy. Recycled newspaper makes for the best eco-friendly pencils (colored or not) which are great for two reasons: 1) they reduce landfill waste by putting newspaper to a second use; 2) they don't promote deforestation.

The tightly coiled newspaper held together by earth safe adhesive is sturdy to hold, while the non-toxic colored graphite goes on smooth and extra dark (which they claim provides double the life of traditional colored pencils).

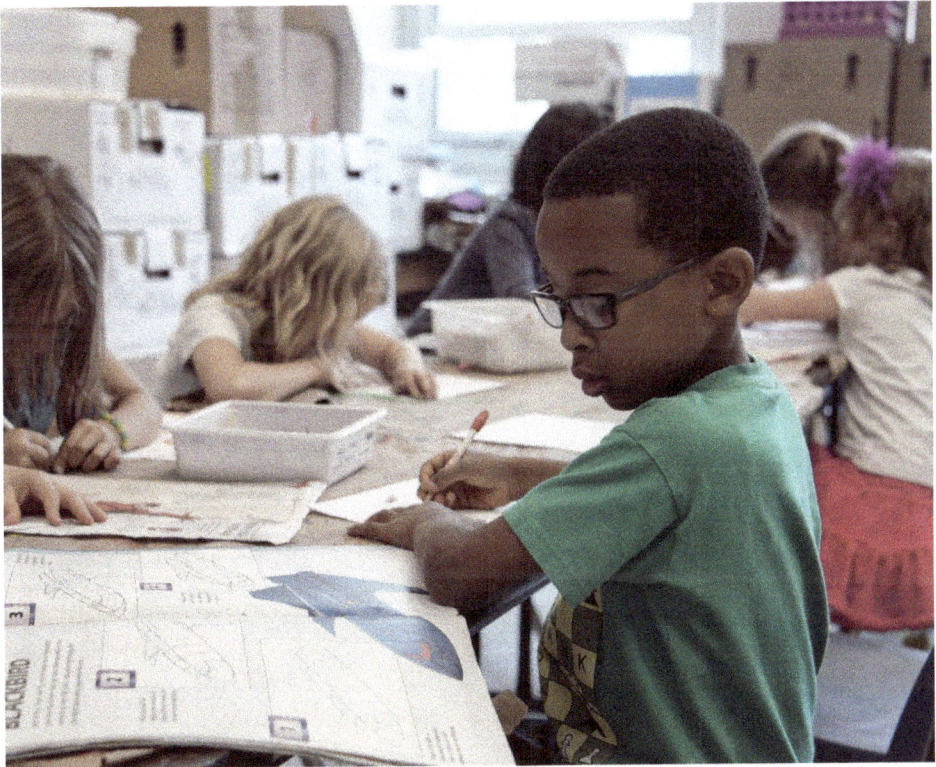

18. Zero Waste Crayons

Conventional crayons are made of petroleum-based paraffin and are tinted with chemical dyes. To that, we say 'time to color outside the lines'. Earth Grown crayons are made of organic, vegan soy wax and mineral pigments that have been certified non-toxic by related third party organizations. All ingredients are sourced from farms and have been grown without pesticides or herbicides. They're also packaged in an uncoated cardboard box with shredded cardboard filling which is 100% compostable.

19. Art Drawing Pad

The Eco Art drawing pad is comprised of FSC and Rainforest Alliance-certified post-consumer agricultural waste. Specifically, it's made from pinzotes, the discarded stalks from banana trees. Each sheet is solid enough to be suitable for paints, markers, and crayons alike.

Environmental Shopping Guide for Paper Products

Sustainable purchasing is about including social, environmental, financial and performance factors in a systematic way. It involves thinking about the reasons for using the product (the service) and assessing how these services could be best met. If a product is needed, sustainable purchasing involves considering how products are made, what they are made of, where they come from and how they will be used and disposed.

Wherever possible RECOMENDED products that employ a combination of characteristics listed in the left hand column, and NOT-RECOMENDED products that demonstrate characteristic in the right-hand column.

RECOMENDED	NOT-RECOMENDED
• International Ecolabel certified • High post-consumer recycled fibre content • Non-wood • Chlorine free • Paper that is less bright	• Not Ecolabel certified • Long distance transport • Unsustainably harvested wood resources

A Simple Sustainable Plan

A	Reduce paper use
B	Use EcoLabel Certified Products
C	Choose High Post-Consumer Recycled Content
D	Choose Non-Wood (Tree-free) Fibres
E	Choose Less Transportation
F	Use Sustainably Harvested Wood Fibre
G	Choose Chlorine-Free Paper
H	Select Paper with Appropriate Brightness

A: Reduce paper use

The first step toward a more sustainable paper cycle is to reduce use. Using less paper saves money and contributes to sustainability by mitigating the environmental impact of production and use. The average North American now uses 227 Kilograms of paper per year, more than double the global average. Consuming less paper reduces the impacts of paper production and the associated energy use from operating printers and copiers. Both of these processes are known to have negative impacts for sustainability and should be minimized.

Strategies for reducing paper use include:
- Electronically archiving instead of printing non-critical documents
- Sharing and reviewing document drafts electronically
- Purchasing a duplex printer/photocopier and selecting double-sided printing as the default
- Re-using paper that is already printed on one side for draft copies

Even if paper use declines in industrialized countries, developing nations will continue to increase their consumption as they gain access to more information and technology. Many countries currently have insufficient paper to fulfill basic education needs. It is therefore essential that paper consumption be closely monitored and restrained so that the resource, and the benefits it provides, can be more equally distributed to meet future needs.

B: Use EcoLabel Certified Products

Environmental Choice certified (eg. FSC certified Vol.6, Page: 29) paper products have met the required standards regarding noxious emissions to water, wastewater discharge levels, use of recycled content, solid waste volume, potential contribution to acid rain and climate change, energy use and forestry and habitat conservation. This widely respected sustainability rating provides an easy way of distinguishing genuinely green products from their competitors.

Courtsey of FSC International (www.FSC.org)

C: Choose High Post-Consumer Recycled Content

As opposed to virgin fibres, post-consumer recycled fibres have been recovered from paper products already "consumed" by an end user. The use of these recycled fibres directly reduces the use of forest resources, in turn mitigating the associated habitat destruction, loss of top soil and other forms of ecosystem damage.

> **Recycling one tonne of paper:**
> • Saves up to 31 trees, 4,000 kWh of energy, 1.7 barrels (270 litres) of oil, 10.2 million Btu's of energy, 26,000 litres of water and 3.5 cubic metres of landfill space
>
> • Burning that same tonne of paper would generate about 750 kilograms of carbon dioxide
>
> • Recycling paper saves 65% of the energy needed to make new paper and also reduces water pollution by 35% and air pollution by 74%

D: Choose Non-Wood (Tree-free) Fibres

Non-wood plant fibres do not need to be bleached with chlorine to be lightened, consume less energy when being processed, release fewer greenhouse gases in the production process and has less harmful water discharge. Crops grown specifically for the purpose of paper production can include Kanaf, jute, flax and hemp. Certain agricultural residues, such as wheat stalks and sugar cane bagasse, can also be processed into non-wood paper. When choosing non-wood fibres, preference should be given to those that are organically and sustainably grown. This eliminates the use of synthetic fertilizers, herbicides and pesticides, reducing the associated ecological and human health impacts.

E: Choose Less Transportation

The proximity of where the fibres are harvested, where the final products are produced and your own location has a significant impact on the sustainability of paper production. Growing, producing and buying locally will reduce emissions from fossil fuels. Transportation may have to be weighed against some of the other desired characteristics. More information regarding sustainable methods of transportation is available in the Transportation Guide.

F: Use Sustainably Harvested Wood Fibre

When using wood-based fibres, it is important to consider how the forest resources from which the paper was derived are managed and harvested. Preference should be given to companies that practice sustainable forestry techniques. Third party organizations such as the Forest Stewardship Council (FSC, Vol.6 Page 29) certify the harvesting and management of forestry resources to ensure long-term sustainability.

G: Choose Chlorine-Free Paper

To reduce the potential risks associated with chlorine compounds, a number of paper manufacturers are switching to chlorine-free compounds for whitening paper. Alternative bleaching agents include: oxygen, hydrogen peroxide or ozone treatments. Paper products often identify the bleaching method used for processing pulp. Paper products processed with derivatives of chlorine produce fewer dioxins than regular chlorine. This process is described as elemental chlorine free (ECF). Products bleached with no chlorine and no chlorine derivatives are sometimes referred to as totally chlorine free (TCF) or process chlorine free.

H: Select Paper with Appropriate Brightness

The brightness of paper is largely a function of the chemicals used in the pulp and/or the amount of recycled fibres used in the paper. Selecting less bright paper can reduce overall impacts. The function or use of the paper influences how bright the paper needs to be. For example, copy paper is generally not used for publicity or advertising. When the content, rather than the appearance of the paper matters, the whiteness is irrelevant as long as the text is legible. It is important to appropriately match the paper to its purpose.

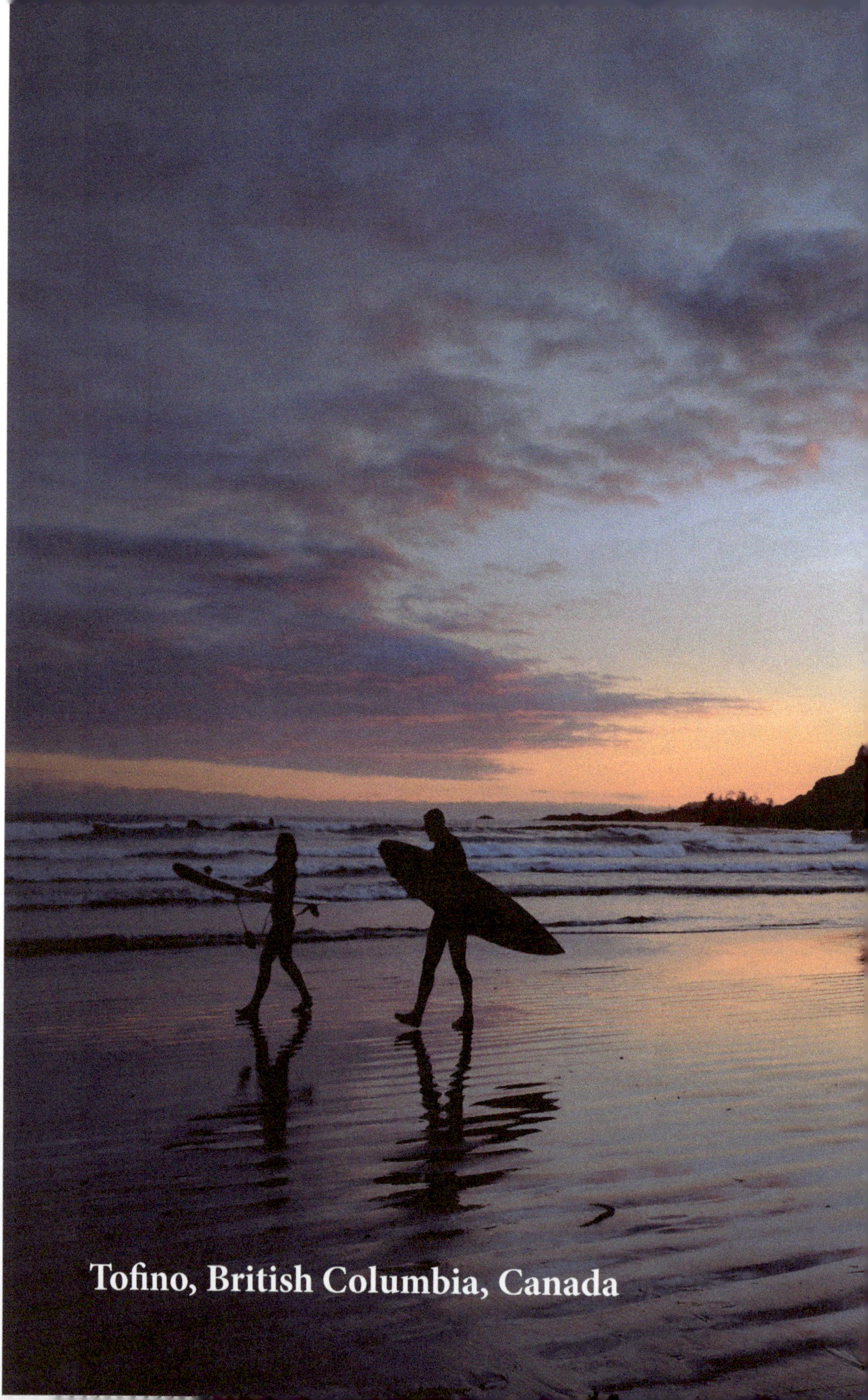

Tofino, British Columbia, Canada

Consumers are increasingly demanding sustainable products that are not toxic to themselves or the environment. The natural market is growing exponentially, and choosing raw, natural materials will cement your brand as a safe choice — both environmentally and economically

CHAPTER **9**

Top Ten Award International Network Environmental Pioneers

Top Ten Award international Network (TTAIN) was established in 2012 to recognize outstanding individuals, groups, companies, organizations representing the best in the public works profession. TTAIN publishing books related to international Eco-labeling plans to increase public knowledge in purchasing based on the environmental impacts of products. We introduce in each volume some of the organizations that are doing their best in relation to taking care of the environmnet.

Eco-Schools

Global

Recognised by UNESCO and UN Environment as a world-leader within the field of Education for Sustainable Development, the Eco-Schools programme is one of five programmes run by the Foundation for Environmental Education (FEE). Based on a 7-Step Methodology, the Eco-Schools programme encourages young people to engage in their environment through project-based, experiential learning, focused on positive sustainable actions. With over 59,000 schools in 72 countries, Eco-Schools is the largest global sustainable schools programme and has helped shaped millions of young people into sustainably-minded, environmentally conscious individuals for more than 25 years.

Contact:
Foundation for Environmental Education
Scandiagade 13,
2450 Copenhagen SV, DENMARK
Website: www.ecoschools.global
E: info@fee.global

Note:
We've done our absolute best to provide the best information possible, but since we haven't tried every single one of these solutions in every possible cleaning situation, we can't vouch for them 100 percent.

UN environment programme

UNEP

The United Nations Environment Programme (UNEP) is the leading global environmental authority that sets the global environmental agenda, promotes the coherent implementation of the environmental dimension of sustainable development within the United Nations system, and serves as an authoritative advocate for the global environment.

Our mission is to provide leadership and encourage partnership in caring for the environment by inspiring, informing, and enabling nations and peoples to improve their quality of life without compromising that of future generations.

Headquartered in Nairobi, Kenya, we work through our divisions as well as our regional, liaison and out-posted offices and a growing network of collaborating centres of excellence. We also host several environmental conventions, secretariats and inter-agency coordinating bodies. UN Environment is led by our Executive Director.

We categorize our work into seven broad thematic areas: climate change, disasters and conflicts, ecosystem management, environmental governance, chemicals and waste, resource efficiency, and environment under review. In all of our work, we maintain our overarching commitment to sustainability.

Website: www.unep.org

Global

The Foundation for Environmental Education (FEE) is the world's largest environmental education organisation, with 100 member organisations in 82 countries. Our educational programmes, Eco-Schools, Learning About Forests and Young Reporters for the Environment, empower young people to create an environmentally conscious world through a solutions-based approach. Our Green Key and Blue Flag programmes are globally recognized for promoting sustainable business practices and the protection of natural resources. With 40 years of impactful experience in ESD, FEE's Strategic Plan, GAIA 20:30, prioritises climate action across all five programmes to address the urgent threats of climate change, biodiversity loss and environmental pollution.

Contact:
Foundation for Environmental Education
Scandiagade 13,
2450 Copenhagen SV, DENMARK
Website: www.fee.global
E: info@fee.global

Germany

FSC® is a global not-for-profit organization that sets the standards for responsibly managed forests, both environmentally and socially. When timber leaves an FSC certified forest they ensure companies along the supply chain meet our best practice standards also, so that when a product bears the FSC logo, you can be sure it's been made from responsible sources. In this way, FSC certification helps forests remain thriving environments for generations to come, by helping you make ethical and responsible choices at your local supermarket, bookstore, furniture retailer, and beyond. www.fsc.org

FSC® International
Adenauerallee 134
53113 Bonn
E-mail: info@fsc.org
Phone: +49 (0) 228 367 66

FSC Canada
50 rue Sainte-Catherine Ouest,
bureau 380B, Montreal, QC H2X 3V4
Email: info@ca.fsc.org
Telephone: 514-394-1137

Bibliography:

Amberg, N.; Magda, R. Environmental Pollution and Sustainability or the Impact of the Environmentally Conscious Measures of International Cosmetic Companies on Purchasing Organic Cosmetics. Visegrad J. Bioecon. Sustain. Dev. 2018, 1, 23.

Asadi, J., "International Environmental Labelling, Economic Consequencies, Export Magazine, July 2001

Asadi, J. 2008. Mobile Phone as management systems tools, ISO Magazine, Vol.8, No.1

Asadi, J., Eco-Labelling Standards, National Standard Magazine, Sep. 2004.

Barbieux, D.; Padula, A.D. Paths and Challenges of New Technologies: The Case of Nanotechnology-Based Cosmetics Development in Brazil. Adm. Sci. 2018, 8, 16.

Advanced Engineering and Applied Sciences: An International Journal 2014; 4(3): 26-28

Berolzheimer, C. (2006). Pencils: An Environmental Profile.

Chemical Week, 1999. Europe's Beef Ban Tests Precautionary Principle. (August 11).

Chaudri, S.K.; Jain, N.K. History of Cosmetics. Asian J. Pharm. 2009, 7–9, 164–167.

CHOI, J.P. Brand Extension as Informational Leverage. Review of Eco- nomic Studies, Vol. 65 (1998), pp. 655-669.

Conway, G. 2000. Genetically modified crops: risks and promise.

Corrado, M., (1989), The Greening Consumer in Britain, MORI, London

Corrado, M., (1997), Green Behaviour – Sustainable Trends, Sustainable Lives?, MORI, london, accessed via countries. Manila, Asian Development Bank 33p.

Davies, Clive. Chief, Design for the Environment Program, Environmental Protection Agency. Interview. March 24, 2009.

Federal Trade Commission, "Sorting Out Green Advertising Claims." http://www.ftc.gov/bcp/edu/pubs/consumer/general/gen02.shtm (March 26, 2009, March 27, 2009)

Ooyen, Carla. Research Manager with Nutrition Business Journal. Personal correspondence. March 19, 2009.

Tekin, Jenn. Marketing Manager with Packaged Facts & SBI. Personal correspondence. March 17, 2009.

University of California - Berkeley. http://berkeley.edu/news/media/releases/2006/05/22_householdchemicals.shtml (March 26, 2009)

U.S. Department of Health and Human Services, Household Products Database.http://householdproducts.nlm.nih.gov/cgi-bin/household/prodtree?prodcat=Inside+the+Home (March 17,

Women's Voices of the Earth, "Household Cleaning Products and Effects on Human Health."http://www.womenandenvironment.org/campaignsandprograms/SafeCleaning/safecleaninghealth (March 17, 2009)

EMONS, W. Credence Goods Monopolists. International Journal of In- dustrial Organization, Vol. 19 (2001), pp. 375-389.

European Union official website: https://ec.europa.eu/info/about-european-commission/contact_en

Feenstra, R.C. "Exact Hedonic Price Indexes," Review of Economics and Statistics 77 (1995): 634-653.

Feenstra, R.C., and J.A. Levinsohn. "Estimating Markups and Market Conduct with Multidimensional Product Attributes," Review of Economic Studies (62 (1995): 19-52.

ForestEthics. (n.d.). Back to School Report Card.

Forest Stewardship Council: "Principles and criteria for forest stewardship" Document 1.2: <http://www.fscoax.org>

Forsyth, K. 1999. Will consumers pay more for certified wood products? Journal of Forestry 97 (2) : 18-22.

ForestChoice #2 (2014, January 1). ForestChoice #2 Graphite Pencils (12 Pack).

Francois, C., Harris, B. (2014, November 2). How are Mechanical Pencils Made?.

Freeman, A. M III. The Measurement of Environmental and Resource Values. Theory and Methods. Washington D.C.: Resource for the Future, 1993.

Friends of the Earth, 1993. Timber certification and eco-labeling. London, FOE:

Geetha Margret Soundri, "Ecofriendly Antimicrobial Finishing of Textiles Using Natural Extract", Journal of International Academic Research For Multidisciplinary, ISSN: 2320 – 5083, 2014, Vol 2.

Graves, P., J.C. Murdoch, M.A. Thayer, and D. Waldman. "The Robustness of Hedonic Price Estimation: Urban Air Quality," Land Economics 64(1988): 220-233.

Halvorsen, R. and R. Palmquist. "The Interpretation of Dummy Variables in Semilogarithmic Equations." American Economic Review 70:474-75 (1980).

Henderson D. (2008). Opportunity Cost." The Concise Encyclopedia of Economics.

How It's Made. (2009, Nov 17). How It's Made Graphite Pencil Leads [video file].

Imhoff, Dan. "Growing Pains: Organic Cotton Tests the Fibre of Growers and Manufacturers Alike," reprinted on Simple Life's web page (simplelife.com), but first printed by Farmer to Farmer, December 1995.

Incomplete Consumer Information in Laboratory Markets. Journal of Environmental labeling.

ISO 14020, ISO 14021,ISO 14024,ISO 14025, International Organization for Standardization.

Kennedy, P.E. "Estimation with Correctly Interpreted Dummy Variables in Semilogarithmic Equations," American Economic Review 71: 801 (1981).

Kirchho®, S., (2000), Green Business and Blue Angels.

Kraus, Jeff. Lab Technician at the North Carolina School of Textiles.

Labeling Issues, Policies and Practices Worldwide.

Lamport, L. 1998. The cast of (timber) certifiers: who are they? International J. Ecoforestry 11(4): 118-122.

Large Scale impoverishment of Amazonian forests by logging and fire. 1999.

Lathrop, K.W. and Centner, T.J. 1998. Eco-labeling and ISO 14000: An analysis of US regulatory systems and issues concerning adoption of type II standards. Environmental

Lee, J. et al. 1996. Trade related environmental measures; sizing and comparing impacts.

Lehtonen, Markku. 1997. Criteria in Environmental Labeling: A comparative Analysis on Environmental Criteria in Selected Labeling Schemes. Geneva, UNEP. 148p.

LIEBI, T. Trusting Labels: A Matter of Numbers? Working Paper Uni versity of Bern, No. 0201 (2002).

Lindstrom, T. 1999. Forest Certification: The View from Europe's NIPFs. Journal of Forestry 97(3): 25-31. London

Losey, J.E., Rayor, L.S. & Carter, M.E. 1999. Transgenic pollen harms monarch larvae. Nature 399 20 May): p.214.

Mattel Ever After High Cedar Wood Doll. (2014, July 3).

Management 22 (2) : 163-172.

Mattoo, A. and H. V. Singh, (1994), Eco-Labelling: Policy Considera-Michaels, R. G., and V. K. Smith. "Market Segmentation And Valuing Amenities With Hedonic Models: The Case Of Hazardous Waste Sites," Journal of Urban Economics, 1990 28(2), 223-242.

Nicholson-Lord, D., (1993) 'Tis the Season to be Green, The Independent, 20 December

Nuttall, N., (1993), Shoppers can cross green products off their lists, The Times, 3 July

OCDE/GD(97)105. Paris, OECD. 81p.

OECD. "Ec-labelling: Actual Effects of Selected Programmes," OCDE/GD (97) 105, 1997, Paris. (available on line at http://www.oecd.org/env/eco/books.htm#trademono)

OECD. 1997a. Case study on eco-labeling schemes. Paris, OECD (30 Dec):

OECD. 1997b. Eco-labeling: Actual Effects of Selected Programs.

Osborne, L. "Market Structure, Hedonic Models, and the Valuation of Environmental Amenities." Unpublished Ph.D. dissertation. North Carolina State University, 1995.

Osborne, L., and V. K. Smith. "Environmental Amenities, Product Differentiation, and market Power," Mimeo, 1997.

Ozanne, L.K. and Vlosky, R.P. 1996. Wood products environmental certification: the United States perspective". Forestry Chronicle 72 (2) : 157-165.

Palmquist, R. B., F. M. Roka, and T.Vukina. "Hog Operations, Environmental Effects, and Residential Property Values," Land Economics 73(1), (1997): 114-24.

Palmquist, R.B. "Hedonic Methods," in J.B Braden and C.D. Kolstad, eds. Measuring the Demand for Environmental Improvement. Amsterdam, NL: Elsevier, 1991.

Paper Mate. (2014). Paper Mate Recycled.

Pento, T. 1997. Implementation of Public Green Procurement Programs (22-31) in Greener Purchasing: Opportunities and Innovations. Sheffield, Greenleaf Publ. 325 p.

Perloff, J. "Industrial Organization Lecture Notes," Mimeo. University of California at Berkeley (1985).

Plant, C. and Plant, J. 1991. Green business: hope or hoax? Philadelphia, New Society Publishers 136 p.

Pencil Making Today (2014, January 1). Pencil Making Today: How to Make a Pencil in 10 Steps.

Polak, J. and Bergholm, K. 1997. Eco-labeling and trade: a cooperative approach (Jan.): Policy in a Green Market. Environmental and Resource Economics 22, 419-

Poore, M.E.D. et al. 1989. No timber without trees. London, Earthscan. 352p.

Raff, D. M.G., and M. Trajtenberg. "Quality-Adjusted Prices for the American Automobile Industry: 1906-1940." NBER Working Paper Series, Working Paper No. 5035, February 1995.

Roberts, J. T. 1998. Emerging global environment standards: prospects and perils. Journal of Developing Societies 14 (1): 144-163.

Rosen, S., "Hedonic Prices and Implicit Markets: Product Differentiation in Pure Competition." Journal of Political Economy. 82: 34-55 (1974).

Ross, B. 1997. Eco-friendly procurement training course for UN HCR. : 126 p.

Ryan, S., and Skipworth, M., (1993), Consumers turn their backs on green revolution, The Times, 4 April

Salzman, J. 1997. Informing the Green Consumer: The Debate over the Use and Abuse of Environmental Labels. Journal of Industrial Ecology 1 (2): 11-22.

Sanders, W. 1997. Environmentally Preferable Purchasing: The US Experience (946-960) in Greener Purchasing: Opportunities and Innovations. Sheffield, Greenleaf Publ. 325p.

Sayre, D. 1996. Inside ISO 14000: The competitive advantage of environmental management. Delray Beach FL., St. Lucie Press. 232p.

Suzuki, D. (2014, January 1). PEG Compounds and their contaminants

SHAPIRO, C. Premiums for High Quality Products as Returns to Reputa- tion. Quarterly Journal of Economics, Vol. 98, No. 4 (1983), pp. 659-680.

Stillwell, M. and van Dyke, B. 1999. An activists handbook on genetically modified organisms and the WTO. Washington DC., The Consumer's Choice Council: 20 p.

Semenzato, A.; Costantini, A.; Meloni, M.; Maramaldi, G.; Meneghin, M.; Baratto, G. Formulating O/W Emulsions with Plant-Based Actives: A Stability Challenge for an Eective Product. Cosmetics 2018, 5, 59.

Sources of Plastics (2014, January 1). Sources of Plastics.

Singh, S. (2008, March 6). Paraffin wax.

Saint Jean Carbon. (n.d.). Sri Lankan Graphite.

Teisl, M. F., B. Roe, and R. L. Hicks. "Can Eco-labels tune a market? Evidence from dolphin-safe labeling," Presented paper at the 1997 American Agricultural Economics Association Meetings, Toronto.

Tollefson, Jennifer E. (2008). Calocedrus Decurrens.

THE GERSEN, C. Psychological Determinants of Paying Attention to Eco- Labels in Purchase Decisions: Model Development and Multinational Vali- dation. Journal of Consumer Policy, Vol. 23, No. 4 (2000), pp. 285-313.

Tibor, T. and Feldman, I. 1995. ISO 14000: a guide to the new environmental management standards. Burr Ridge Ill., Irwin Professional Publ. 250 p.

TU.S. Energy Information Administration, What is U.S. Electricity Generation by Energy Source?, Retrieved From: https://www.eia.gov/tools/faqs/faq.php?id=427&t=3

U.S. Energy Information Administration, Biomass Explained, Retrieved From: https://www.eia.gov/energyexplained/?page=biomass_home

U.S. Environmental Protection Agency. National Water Quality Fact Inventory: 1990 Report to Congress. EPA 503-9-92-006, Apr. 1992.

UK Eco-labelling Board website, accessed via http://www.ecosite.co.uk/Ecolabel-UK/

US Environmental Protection Agency (EPA742-R-99-001): 40 p. <www.epa.gov/opptintr/epp>

US EPA, 1993. Determinants of effectiveness for environmental certification and labeling programs. Washington, D.C., US Environmental Protect

US EPA, 1993. Status report on the use of environmental labels worldwide. Washington, D.C., US Environmental Protection Agency (742-R-93-001 September).

US EPA, 1993. The use of life-cycle assessment in environmental labeling. Washington, D.C., US Environmental Protection Agency (742-R-93-003 September).

US EPA, 1998. Environmental labeling: issues, policies, and practices worldwide. Washington DC., Environmental Protection Agency, Pollution Prevention Division Prepared by Abt

US EPA, 1999. Comprehensive procurement guidelines (CPG) program. Washington, D.C., US Environmental Protection Agency: <www.epa.gov/cpg>

US EPA, 1999. Environmentally preferable purchasing program: Private sector pioneers: How companies are incorporating environmentally preferable purchases. Washington University of Saskatchewan, Sustainable purchasing guide.

USG, 1993. Federal acquisition, recycling, and waste prevention. Washington DC., Executive Order: (20 October).

USG, 1998. Greening the government through waste prevention, recycling, and federal acquisition. Washington, D.C., Executive Order 13101 (September).

Kijjoa, A.; Sawangwong, P. Drugs and Cosmetics from the Sea. Mar. Drugs 2004, 2, 73–82. [CrossRef]

Wang, J.; Pan, L.; Wu, S.; Lu, L.; Xu, Y.; Zhu, Y.; Guo, M.; Zhuang, S. Recent Advances on Endocrine Disrupting Eects of UV Filters. Int. J. Environ. Res. Public Health 2016, 13, 782.

Bilal, A.I.; Tilahun, Z.; Shimels, T.; Gelan, Y.B.; Osman, E.D. Cosmetics Utilization Practice in Jigjiga Town, Eastern Ethiopia: A Community Based Cross-Sectional Study. Cosmetics 2016, 3, 40.

Ting, C.T.; Hsieh, C.M.; Chang, H.-P.; Chen, H.-S. Environmental Consciousness and Green Customer Behavior: The Moderating Roles of Incentive Mechanisms. Sustainability 2019, 11, 819.

Chen, K.; Deng, T. Research on the Green Purchase Intentions from the Perspective of Product Knowledge. Sustainability 2016, 8, 943.

Wang, H.; Ma, B.; Bai, R. How Does Green Product Knowledge Eectively Promote Green Purchase Intention? Sustainability 2019, 11, 1193.

Nguyen, T.T.H.; Yang, Z.; Nguyen, N.; Johnson, L.W.; Cao, T.K. Greenwash and Green Purchase Intention: The Mediating Role of Green Skepticism. Sustainability 2019, 11, 2653.

Cinelli, P.; Coltelli, M.B.; Signori, F.; Morganti, P.; Lazzeri, A. Cosmetic Packaging to Save the Environment: Future Perspectives. Cosmetics 2019, 6, 26.

Eixarch, H.; Wyness, L.; Siband, M. The Regulation of Personalized Cosmetics in the EU. Cosmetics 2019, 6, 29.

APPENDIX I: SEARCH BY LOGOS

Here you can search the logos in this volume. It will help you to better undersand the Ecolabels you may encounter while shopping. Buying Eco-products will aid in having a better environment with minimum polution during production processes. Three important parameteres for shopping are **quality**, **price** & **environmental impacts** of the products.

Vol.6 Goto page: 31	Vol.6 Goto page: 79
Vol.6 Goto page: 50	Vol.6 Goto page: 44
Vol.6 Goto page: 39	Vol.6 Goto page: 83
Vol.6 Goto page: 29	Vol.6 Goto page: 43

BIO LATINA CERTIFICADORA	Korea Eco-Label
Vol.6 Goto page: 35	Vol.6 Goto page: 37
	FSC FORESTS FOR ALL FOREVER ™
Vol.6 Goto page: 50	Vol.6 Goto page: 29
FSC FORESTS FOR ALL FOREVER ™	CEC
Vol.6 Goto page: 29	Vol.6 Goto page: 32
NORDIC SWAN ECOLABEL	ฉลากเขียว GREEN LABEL : THAILAND
Vol.6 Goto page: 42 ,41	Vol.6 Goto page: 43

Vol.6 Goto page: 30	Vol.6 Goto page: 36
Vol.6 Goto page: 40	Vol.6 Goto page: 34
Vol.6 Goto page: 31	Vol.6 Goto page: 38
Vol.6 Goto page: 33	Vol.6 Goto page: 29

Appendix II

PAPER MADE OUT OF ALGAE CELLULOSE, A SUSTAINABLE ALTERNATIVE TO CONVENTIONAL INDUSTRY

The main source of cellulose has always been wood from trees and other vascular plants. The cellulose obtained from these sources is associated with other natural polymers, mainly lignin. Algae can be considered as an alternative source of cellulose to traditional raw materials. One of the principal problems regarding the conventional extraction of cellulose is the removal of lignin. The lignin content in the algae cell wall is so low that there are not problems associated with lignin removal. Two of the most important bloom forming kinds of algae, Ulva sp. and Cladophora sp., as a raw material for papermaking. The amount of solvent-substances, lignin and holocellulose in dried algae pulp is estimated. The results show that the studied algae have low lignin-like compounds and solvent-soluble substances content, which supposes an enormous advantage over the current cellulose extraction methods as it eliminates the need of pre-treatment, cooking and bleaching stages. Therefore, the application of extremely tox-

ics reagents used nowadays is not necessary. The holocellulose content obtained , ranged from 47 to 54%, lower than that of wood or herbaceous species.

The algae genus stand as a proper source of reinforcing fibers for papermaking purposes. Also, represent an excellent opportunity to valorize tidal wastes obtained from bloom episodes.

Environmental Friendly Photos

Environmental friendly photos will be placed in this appendix. These photos can be received in the Top Ten Award International Network inbox from anywhere and everywhere, all over the globe. You can send your appropriate photos to us for them to be considered for publishing in one of the future, related volumes. They will be published with proper credit to the sender. The pictures can also be images of the Ecolabels existing in products within your country.

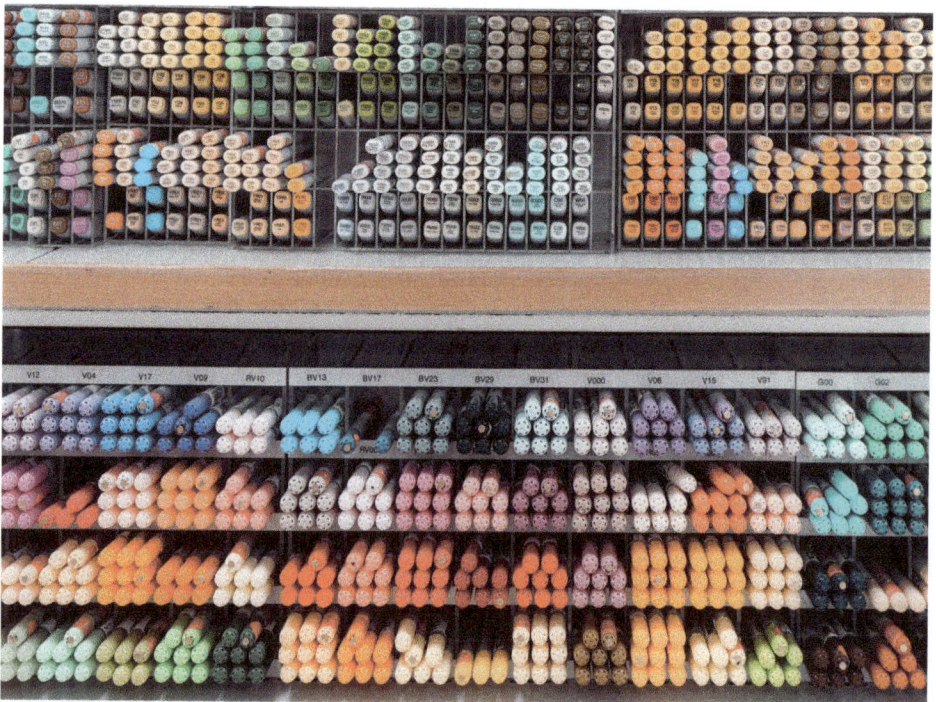

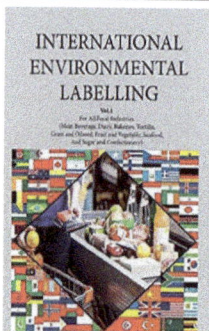

INTERNATIONAL ENVIRONMENTAL LABELLING	# Vol.1 For All People who wish to take care of Climate Change, Food Industries: (Meat, Beverage, Dairy, Bakeries, Tortilla, Grain and Oilseed, Fruit and Vegetable, Seafood, And Sugar and Confectionery)
INTERNATIONAL ENVIRONMENTAL LABELLING	# Vol.2 For All People who wish to take care of Climate Change, Electrical Industries: (Renewable Energy, Biofuels, Solar Heating & Cooling, Hydroelectric Power, Solar Power, Wind Power, Energy Conservation, Geothermal and Nuclear Power)
INTERNATIONAL ENVIRONMENTAL LABELLING	# Vol.3 For All People who wish to take care of Climate Change, Fashion & Textile Industries: (Fashion Design, The Fashion System, Fashion Retailing, Marketing and Marchandizing, Textile Design and Production, Clothing and Textile Recycling)
INTERNATIONAL ENVIRONMENTAL LABELLING	# Vol.4 For All People who wish to take care of Climate Change, Health & Beauty Industries: (Fragrances, Makeup, Cosmetics, Personal Care, Sunscreen, Toothpaste, Bathing, Nailcare & Shaving, Skin Care, Foot Care, Hair Care and Other Health & Beauty Products)

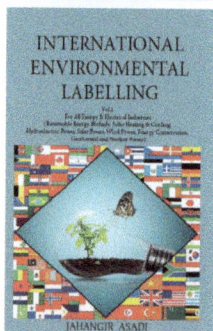

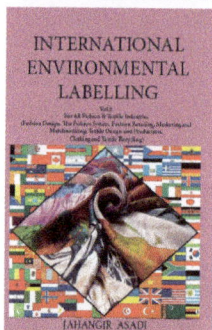

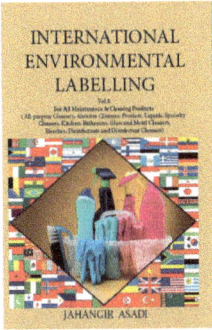

INTERNATIONAL ENVIRONMENTAL LABELLING	**Vol.5**
	For All People who wish to take care of Climate Change, Maintenance & Cleaning Products: (All-purpose Cleaners, Abrasive Cleaners, Powders. Liquids, Specialty Cleaners, Kitchen, Bathroom, Glass and Metal Cleaners, Bleaches, Disinfectants and Disinfectant Cleaners)
INTERNATIONAL ENVIRONMENTAL LABELLING	**Vol.6**
	For All People who wish to take care of Climate Change, Wood & Stationery Industries: (Wooden Products, Cardboard, Papers, Markers, Pens, NoteBooks, Writing Pads and Writing Sets, Pencils, White Papers, Envelopes and Organizers, Staplers and Paper Clips)
INTERNATIONAL ENVIRONMENTAL LABELLING	**Vol.7**
	For All People who wish to take care of Climate Change, DIY & Construction Industries: (Do it yourself " ("DIY") of Building, Modifying, or Repairing, Renovation, Construction Materials, Cement, Coarse Aggregates. Clay Bricks, Power Cables, Pipes and Fittings, Plywood, Tiles, Natural Flooring)
INTERNATIONAL ENVIRONMENTAL LABELLING	**Vol.8**
	For All People who wish to take care of Climate Change, Agricuture & Gardening Industries: (Shifting Cultivation, Nomadic Herding, Livestock Ranching, Commercial Plantations, Mixed Farming, Horticulture, Butterfly Gardens, Container Gardening, Demonstration Gardens, Organic Gardening)

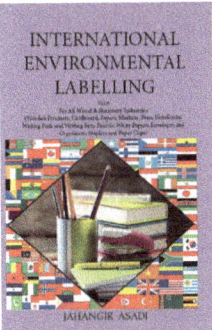

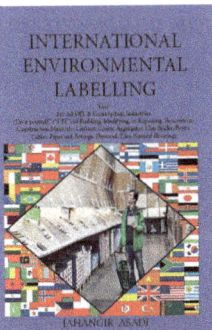

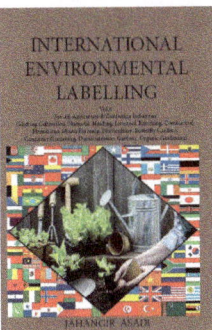

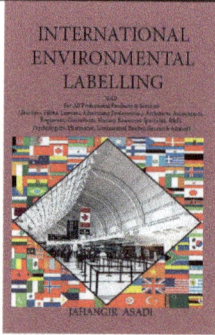

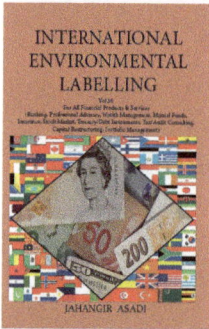

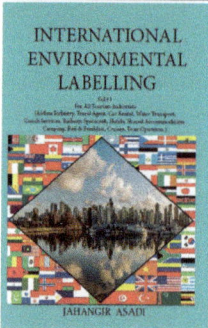

www.ingramcontent.com/pod-product-compliance
Lightning Source LLC
Chambersburg PA
CBHW040757220326
41597CB00029BB/4974